Jose Henrique Bizinoto
Monica Cassia S. Martines

Solving Spatial Metric Geometry Problems

Jose Henrique Bizinoto
Monica Cassia S. Martines

Solving Spatial Metric Geometry Problems

using Information and Communication Technology

ScienciaScripts

Imprint

Any brand names and product names mentioned in this book are subject to trademark, brand or patent protection and are trademarks or registered trademarks of their respective holders. The use of brand names, product names, common names, trade names, product descriptions etc. even without a particular marking in this work is in no way to be construed to mean that such names may be regarded as unrestricted in respect of trademark and brand protection legislation and could thus be used by anyone.

Cover image: www.ingimage.com

This book is a translation from the original published under ISBN 978-613-9-61386-1.

Publisher:
Sciencia Scripts
is a trademark of
Dodo Books Indian Ocean Ltd. and OmniScriptum S.R.L publishing group

120 High Road, East Finchley, London, N2 9ED, United Kingdom
Str. Armeneasca 28/1, office 1, Chisinau MD-2012, Republic of Moldova, Europe
Printed at: see last page
ISBN: 978-620-7-61777-7

Table of contents:

THANKS

First of all, I would like to thank God for having illuminated my path, giving me the strength to endure life's adversities, the health to achieve things that would be impossible without him, and the humility to know that I am nothing without him.

To my wife Maria Isabel, whom I love so much, for her companionship, dedication and love. My beloved daughters, Cristielle and Leticia, for their patience when I was away. To my father Herminio and my dearly missed mother Maria Isabel, for working so hard to make me the man I am, and for the sacrifices they made to ensure my studies.

To my advisor, Ménica, for her dedication, commitment, promptness and inspiration in guiding me, without which I would not have been able to complete this work.

To all my teachers, from the time I started school to my master's degree, because without each one of them I wouldn't be the professional I am.

To the friends I've made throughout my life, who encouraged me when adversity struck, for the laughs we had together. Especially to the family I made at PROFMAT.

To the colleagues and friends who have worked with me throughout my professional life, especially my friend Marquinho, who has always supported me.

To all the people who cheered for or against me, who in some way gave me inspiration to walk and be happy.

Summary

The aim of this work is to propose an activity for secondary school students to teach some concepts of Spatial Metric Geometry, specifically calculating the areas and volumes of geometric solids such as cones, cylinders, spheres, pyramids and prisms, with the help of technology, using the GeoGebra *software*. To develop this activity, we looked at some trends in teaching and learning mathematics, focusing mainly on problem-solving and information and communication technology.

Keywords: Problem Solving; Information and Communication Technology; Spatial Metric Geometry.

INTRODUCTION

Those who live in schools on a day-to-day basis are well aware of the new challenges posed by teaching today, especially teachers. They are constantly being provoked to rethink their teaching practice in the face of scenarios within the classroom that seem to be reshaping themselves at an ever-increasing speed.

These changes are not just happening in one environment or another, in this or that school, at this or that level of education. It is well known that schools are being urged to permanently reorganize themselves in the face of these new demands that emerge from the society in which they are inserted, whether *simply* because of the characterization of the generations of students or because of the *complex* demands of the job market and professional qualification.

This study will analyze a small but important slice of the whole conglomerate that makes up education: the teaching of mathematics at basic education level. From this point of view, this research[1] aims to propose a didactic-methodological alternative that includes activities for secondary school classrooms, which can be characterized as an "option" compared to what is already known and practiced.

To this end, this proposal will take into account aspects related to current trends in mathematics education, especially the perspective of problem solving combined with information and communication technologies. In the midst of the theoretical framework, based on a brief examination of the main current trends, this research will focus on these two trends, interweaving them so that they can be understood as a single methodological trend.

Based on the experiences of the researchers who make up this research, who have been classroom teachers for more than 20 years, we have seen a disparity between how teachers propose to teach mathematics and how students are willing to study it. Even though it is practically impossible for there to be perfect harmony between the aims of these two parties.

The driving force behind this research was the possibility of building a methodological alternative that, in order to overcome these differences, could create paths to more meaningful learning based on the appropriation and use of concepts and experiments inherent to the social and cultural reality of the students involved, who in this case are mostly from low-income backgrounds and live in rural areas.

Among the contents of Mathematics, this study will focus on Geometry, which, according to the particular experiences of the researchers, is the one that generates the most opportunities for proposing methodological alternatives and the one that is least understood by students, due to the difficulty of visualizing and connecting figures with their concepts.

In general, this research aims to develop a *list of* activities in order to create possibilities for re-signifying the way Geometry is taught and learned in high school courses, in this case, with activities focused on the reality of a technical course in Agriculture integrated with high school. This course works with two technical strands, Agronomy and Zootechny, with integrated activities for the common core subjects and those of the technical area.

More specifically, the aim is to adapt a didactic-methodological proposal for teaching Geometry to current trends in Mathematics Education in order to facilitate the teaching of Mathematics; to help promote teaching strategies that take into account the contextualization of students' previous knowledge, taking into account, whenever possible, the social and cultural environments in which they are inserted; and, finally, to promote the culture of inserting technologies into teaching, especially the GeoGebra 5 *software.*0 software, free, dynamic geometry and algebra *software,* especially in mathematics, understood as a way of complementing the teacher's resources and possibilities.

The first chapter describes an overview of trends in Mathematics Education, showing the historical evolution, benefits and difficulties of teaching Mathematics using various techniques, as well as a brief history of Mathematics Education in Brazil.

The second chapter discusses the more in-depth aspects of the *Problem Solving* and *Information and Communication Technology* trends, with an overview of their evolution and current applications.

In the third chapter, we will propose a mathematical teaching-learning activity for the second grade of secondary school in a course integrated into the technical course in Agriculture, but which could be suitable for any student in the same grade, in any secondary school. Problem-solving and computer science will be used to calculate the area and volume of the geometric solids prism, pyramid, cylinder, cone and sphere, as well as their planar shapes, with the exception of the sphere.

---

[1] developed from a joint project with Geraldo Henrique Alves Pereira, students on the Professional Master's Degree in Mathematics in the National Network at UFTM.

CHAPTER 1

A vision of the
Mathematics Education

The theoretical discussion that begins will be established according to a chronology of Mathematical Education in Brazil. Although briefly, this chapter will seek to outline the main academic paths that this area has followed since the first discussions until its structuring as a field of study and research.

1.1   Historical introduction

In the 1950s, discussions on ways of teaching and learning mathematics began with greater intensity. According to Soares (2005), these discussions took place at the first National Congresses of Mathematical Education organized in Brazil:

- 1955 in Salvador: this congress aimed to address issues such as programs and curricula, the class book and *modern trends in teaching, as* well as the improvement of mathematics teachers.
- 1957 in Porto Alegre: at the second congress, it was proposed to study issues related to the learning of mathematics at different levels of education; to define the bases for the elaboration of programs *taking into account scientific and psychological aspects,* seeking to establish norms for *a good articulation between the programs of the different levels of education, in* addition to also studying the influence of mathematics on other subjects.
- 1959 in Rio de Janeiro: the basic aim of the third congress was to study problems relating to secondary, primary, commercial, industrial and normal education; issues relating to the training of secondary teachers were also discussed.

After this date, the 1962 and 1966 Congresses took place, where the agenda was the Modern Mathematics Movement, which had taken hold in Brazilian schools at the time (SOARES, 2005). This movement focused only on the question of mathematical language and its formalization. The need for a greater number of scientists and technicians with better qualifications, coupled with the discourse of the inevitability of a minimum modern scientific education in the face of the "technologization" of society, were some of the justifications for this movement (FONSECA, 2012).

According to Onuchic and Allevato (2004, p. 215), this teaching "emphasized many properties, was excessively concerned with mathematical abstractions and used a universal, precise and concise language". Along the same lines, Pinto (2005, p. 2) states that:

> Triggered on an international level, this movement affected not only the aims of teaching, but also the traditional contents of mathematics, giving primary importance to axiomatization, algebraic structures, logic and sets.

However, instead of improving the teaching and learning of mathematics, the problems worsened, as students absorbed complex ideas but did not learn the mathematical concepts. In the 1970s, the Modern Mathematics Movement came under a lot of criticism from French teachers, who by then had already established the Institutes for Research in Mathematics Education - IREM (SOCIEDADE BRASILEIRA DE EDI CACAO MATEMÁTICA, 2015).

As a result of these congresses, circles and associations of Mathematics Teachers and Researchers emerged, which led to the State Congresses of Mathematics Teachers becoming more frequent (SOCIEDADE BRASILEIRA DE EDUCAQAO MATEMÁTICA, 2015).

The participation of eleven Brazilian researchers in the sixth Inter-American Conference on Mathematical Education (CIAEM) in Mexico in 1985 inspired the creation of the National Meetings on Mathematical Education (ENEM). In 1987, the 1st ENEM took place, which was fundamental for the creation of the Brazilian Society of Mathematical Education (SBEM), which was founded during the 2nd ENEM, in 1988, in the city of Maringá-PR (SOUZA, 2005).

The creation of a society like the SBEM made it possible to bring together professionals who, since the 1950s and 1960s, had been fostering important discussions about the mathematics classroom and who therefore needed their own sword in the academies.

Thus, with the creation of the SBEM, the methodological aspects of the classroom were better discussed and organized, and began to play an important role in the development of Mathematics Education. These methodologies have been called trends in mathematics education (SOCIEDADE BRASILEIRA DE EDUCAQAO MATEMÁTICA, 2015).

In time, discussions about Mathematics Education and its trends have been going on in Brazil for a few decades and, among its comings and goings, it was only in the last thirty years that the subject entered universities and became a research area.

1.2 Trends in Mathematics Education: an introduction

Trends in Mathematics Education are teaching and learning techniques that help to bring concepts closer to students.

According to Mendes (2006), states and research into Mathematics Education have sought to offer theoretical and methodological support to overcome the difficulties encountered by teachers and students during the Mathematics education process.

According to Lopes and Borba *apu5* Flemming, Luz and Mello (2005, p. 15),

> A trend is a way of working that has emerged from the search for solutions to problems in mathematical education. As soon as it is used by many teachers or, even if it is little used, results in successful experiences, we are looking at a real trend.

They also point out that Critical Mathematics Education, Ethnomathematics, Mathematical Modeling, the Use of Computers and Mathematical Writing are real trends.

According to Gomes *2* Rodrigues (2014, p. 60),

> it is very important to point out that, in the classroom, teachers end up using many tendencies in a given activity. This is because, often, due to their own academic background, they have been taught a wide variety of trends by their teachers. Teachers can draw on their creative potential to choose activities that feature the use of many trends.

Thus, it can be seen that the experiences of math teachers in today's schools need more reflection, because the school is living in new times, with new concepts, new challenges, requiring new meanings in their work.

When we talk about mathematics, which is intrinsic to this contemporary school, we immediately ask ourselves whether the school of yesteryear has achieved the same results now. However many points of view there are on the success or otherwise of this question, the fact is that something seems to be agreed upon by the entire academic community (not just mathematics), education professionals, parents and others: the teaching of mathematics needs new models.

This *novelty,* at first glance, opposes the apparent immutability of mathematics. However, this discussion does not invade mathematics itself. What is currently being called for is a coherent discussion about how this science should be applied in the classroom, in basic education courses, and how it motivates its main agent: the teacher, whose praxis is subject to various environmental factors. It is therefore necessary for today's mathematics classroom to be rethought and adapted to the new molds of society, the professions and general education.

This is the field of inquiry of Mathematics Education and the area of activity of its researchers who are dedicated to examining various methodologies of the teaching process. Although the study of teaching methodologies is not a field restricted to Mathematics Education, in this area it has found fertile ground and has been consolidated within undergraduate and postgraduate programs. On this basis, methodological tendencies in mathematical teaching have taken pride of place in the publications of the main scientific journals in the country, among which we can highlight Bolema (Boletim de EducaQão Matemática da Universidade Estadual Paulista - Campus Rio Claro) and Zetetiké (Revista de EducaQão Matemática da Faculdade de EducaQáo da Universidade Estadual de Campiñas e Universidade Federal Fluminense), as well as several others from the postgraduate programs themselves.

Over the next few lines, this study will examine what is proposed in high school courses, under the analysis of current methodological trends in teaching mathematics, listed according to their prominence in national academic/scientific production, as described above. With this in mind, the main current methodological trends to be discussed here are:

- Ethnomathematics
- History of Mathematics
- Mathematical games and concrete materials
- Critical Mathematics
- Mathematical modeling
- Problem Solving
- Information and Communication Technologies

Therefore, in addition to the historical circumstances, we will present and discuss the educational concepts and aspects of these trends more specifically in the following sections.

1.2.1   Ethnomathematics

As a technique to bring mathematics closer to students, the study of Ethnomathematics emerged in the 1970s. According to Costa (2014), Ubiratan D'Ambrosio developed this teaching method as a critique of the traditionalism of mathematics teaching, by analyzing applications in various socio-cultural contexts. "The word came from the combination of *techné* (way of doing, technique), *materna* (living with the socio-cultural reality, teaching, explaining) and *eíno* (insertion of man into the cultural environment)" (COSTA, 2014, p. 182).

This methodology takes into account the student's previous knowledge, used in their ethnicity, in their home, in their parents' profession, among others. For Zorzan (2007), the main reason for Ethnomathematics becoming the focus of research is the need to reflect on the importance of valuing cultural knowledge and rebuilding the self-esteem of the people, who also have their riches, values and knowledge. However, it should not be adopted as the only way of teaching mathematics.

Costa (2014) states that Ethnomathematics is not a teaching method, but a plan for inclusive actions between teachers and students, or even a human action in the production of contextualized knowledge, by the different cultural forms in the most diverse human groups.

Ethnomathematics is not opposed to traditional mathematics; both aim to improve knowledge, demonstrate tools for mastering numbers and generate mathematical concepts to be used for their own benefit, but in different ways. In this way, the former introduces mathematical concepts by looking at the concepts acquired; the latter approaches the concepts in a cognitive way, but to be used on a daily basis.

According to Costa (2014, p. 188),

> all the peoples of the world have dedicated themselves to mathematizing their problems, but in order to solve them, and not merely for scientific or instructional purposes. Ethnomathematics can make a decisive contribution to a better understanding of the world, making it more humanized and less technocratic.

But how can ethnomathematics be introduced into mathematics teaching? "Mathematics, as a school subject, needs to be worked on in a textualized way and subject to different relationships with other areas of knowledge and with the needs and life history of the social group" (ZORZAN, 2007, p. 81). This practice, like each of the other teaching trends, should not be used as the only way of teaching, but as one of the tools to arouse students' interest in the teaching-learning process of mathematics. In addition, it should be approached in a timely manner, in various concepts developed by teachers, in the most diverse grades of primary and secondary education, because if interest does not occur early in the student's life, when the desire to learn awakens, they will not have grasped the concepts necessary to be able to use algorithms productively and may give up on the learning process.

For Costa (2014, p. 186):

> Three key stages can be identified when developing an ethnomathematics pedagogy. The first is the investigation stage, when the students are confronted, in a round table process, with the objectives to be achieved, informing them of the precepts that distinguish it from traditional mathematics teaching. In the second stage, that of discussion, the teacher listens to the students about which themes will be organized and developed in the light of their reality.
>
> In the third phase, problematization, the learning situations will focus on activities.

Ethnomathematics is an inclusive mathematical approach, because it works with conceptual approaches, taking into account the students' knowledge, to improve their social lives and so that they can understand the attitudes of the rulers, who use linguistic and mathematical knowledge, as well as others, to dominate the intellectually underprivileged classes. Zorzan (2007, p. 80) comments that "ethnomathematics has a political dimension at its core because, by conceiving mathematics as a cultural product, it makes it a people's science, reclaiming it as a historical subject".

Thus, Ethnomathematics as a methodological trend can be used by teachers as a way of bringing school mathematics closer to everyday mathematics. Another methodological trend that can also help teachers prepare their lessons is the History of Mathematics, which will be presented in the next section.

1.2.2   History of Mathematics

The History of Mathematics as a methodological trend proposes teaching mathematics using, among other things, the historical context. Among its various strands and methodological arguments, it aims to show the student that the concepts studied in the classroom were also needed at another time to solve a particular problem, and that they can therefore be interconnected now with their experience.

With the challenge of learning, generated by the history of these concepts and their interrelationships with reality, students can end up motivating themselves to tackle more complex content and, finding a contextualized result, satisfy themselves more effectively; after all, they know where it came from, what it's for and how/where it will be useful.

Valente (2008) states that those concerned with the teaching of mathematics should not neglect its historical dimension. Although the term "history" can have many meanings within mathematics, in methodological terms the teacher should not associate it with "tales" or "anecdotes" (MIGUEL, 1997).

According to Siqueira (2007, p. 27),

> By understanding how mathematics developed, how it influences other knowledge and is also influenced by it, students will also be able to better understand man's difficulties in developing mathematical ideas. In this way, the History of Mathematics can provide students with a dynamic view of the evolution of mathematics in science, technology and society.

In the same vein, Gomes and Rodrigues (2014, p. 66) comment that:

> When historical concepts are integrated, showing the needs and reasons for their emergence, there is motivation in the classroom, so the teacher can make the student understand that mathematics is a concrete science built on their own temperamental emergencies.

In this way, it is prudent to reinforce the understanding that the History of Mathematics should not be, in itself, a *motivating element,* nor *the problem to be dealt with in* class, nor the link between the two, because, according to Miguel (1997, p. 82),

> the motivating aspect of a problem does not lie in the fact that it is "historical" or even in the fact that it is a "problem", but in the greater or lesser degree of challenge that this problem offers, in the way in which this challenge is perceived by the learner, in the type of relationships that are established between this challenge and the values, interests and skills socially constructed by the learner, etc.

According to the text by Miguel (1997), who by the way analyzes various arguments for using the History of Mathematics very well, it is necessary to approach this methodology for the pedagogical purposes of a classroom. According to the author, it seems

> it is more appropriate to take an intermediate position which believes that history - only when properly reconstructed for explicitly pedagogical purposes and organically articulated with the other variables involved in the didactic planning process - can and should play a subsidiary role in Mathematics Education, namely that of a reference point for pedagogical problematization (MIGUEL, 1997, p. 101).

According to Miguel (1997), a pedagogically oriented methodology like this would be of great help to teachers who want to counteract technical teaching tendencies, but it needs to be built from the point of view of the math educator.

Finally, the History of Mathematics is not intended to elucidate all the points that involve the construction of a given piece of knowledge over time, but, as it is structured as a teaching methodology, it seeks to help answer questions that demystify students' ideas related to the adaptation of this science to the times and, above all, it brings with it motivating aspects for learning. In the next topic we'll look at Mathematical Games and Concrete Materials, which is a methodology that involves playfulness and can teach mathematics without directly applying its concepts.

1.2.3   Mathematical Games and Concrete Materials

Concrete materials are useful for learning mathematics because, by constructing objects, students are faced with the need to learn concepts that are abstract to them. Especially in Geometry, this material should portray the geometric elements in the most ideal way, so that their visualization is not so far removed from the real concept.

For this reason, according to Passos *apud* Murari (2011), the materials to be chosen must (i) provide a true embodiment of the mathematical concept or ideas to be explored; (ii) clearly represent the mathematical concept; (iii) be motivating; (iv) be appropriate for the different years of schooling and the different levels of concepts; (v) form a basis for abstraction; (vi) provide individual manipulation.

Math teaching activities using these materials should be related to the concepts students see in class and linked to objects they come into contact with on a daily basis. They can also be interfaced with historical, physical, geographical and other figures, so that the students are interested in the playful activity and it is not just a time to relax from the teaching and learning process.

In addition, Mendes (2006) reaffirms the importance of progressive learning, which does not end with the manipulation of physical models, but with the manipulative-symbolic and abstractive aspects established in each activity.

In this sense, the manipulation of objects can be a didactic-pedagogical resource to be used in math classes, and we can also use games through this manipulation.

Games can be used as a classroom activity in which students need to develop the spirit of planning, because in order to get the expected result, a strategy needs to be worked out and well executed; when they don't generate expected results, they develop the ability to analyze and correct mistakes.

However, mathematical games and concrete materials cannot be seen as the solution to teaching mathematics. According to Fiorentini and Miorim (1990, p. 3):

> Teachers cannot subjugate their teaching methodology to some kind of material because it is attractive or entertaining. No material is valid on its own. The materials and their use should always be in the background. Simply introducing games or activities into mathematics teaching does not guarantee better learning of this subject.

For these authors, games and materials should be used at the beginning of learning activities, to introduce a concept, or at the end, as a consolidation method.

1.2.4 Critical Mathematics

This trend is a concept based on Paulo Freire's pedagogy[2] with Ethnomathematics, in which the student should be able not only to operate and understand mathematical concepts, but also to reflect and position themselves critically on these concepts and be able to act on them (SOARES, 2008).

Also according to Soares (2008), educators must be able to propose and solve questions, as well as question the answers and the questions they propose. The author believes that math students are not having the ability to reflect on the questions and their answers.

Teachers need to be more democratic in their classroom activities and not act in a decisive and prescriptive way, because in order for students to learn, they need to be part of the teaching-learning process, not just as spectators, but as active elements in this process. Within this context, Siqueira (2007, p. 28) quotes Paulo Freire on the importance of dialog in the classroom:

> Through dialog, the teacher-of-the-students and the students-of-the-teacher break down and a new term emerges: teacher-students with student-teachers. The teacher is no longer merely the one who teaches, but someone who is also taught in dialog with the students, who in turn, while they are teaching, are also learning. They become jointly responsible for a process in which everyone grows.

For Skovsmose (2001), critical education must be developed in a relationship of partnership between the teacher and the students. The subject should interest both of them, with the students being the most interested, because they, in a dialog with the teachers, will define what is relevant to the educational process. As the aim is to develop a critical capacity, this capacity cannot be imposed, but rather developed with the skills of the students.

Skovsmose (2001) considers Critical Mathematics to be mathematical literacy and defends it as a way of freeing human beings from the bonds of society. He adds: "Mathematical literacy can be used for the purpose of 'liberation', because it can have the meaning of organizing and reorganizing interpretations of social institutions, traditions and proposals for political reforms" (SKOVSMOSE, 2001, p. 122).

As its name implies, this trend is a Mathematics Education© in which critique constructs teaching; where students and teachers actively participate in lessons and can achieve significant growth in their academic and personal lives.

1.2.5 Mathematical modeling

Various dictionaries converge the meaning of the word *modeling* to "to elaborate© by model or by a mold"; "to give shape to"; "to mold". These synonyms, applied to mathematics, represent a teaching trend in which mathematical models are used to solve real everyday situations.

According to Gomes and Rodrigues (2014, p. 62), modeling "is a different way of looking at mathematics and consists of the art of turning problems from reality into mathematical problems and solving them by interpreting© their solutions in the language of the real world."

More specifically, Bassanezi (1999, p. 12) builds the norm of this methodology from the initial concept of a *mathematical model,* which "is a consistent set of equations or mathematical structures, elaborated to correspond to some phenomenon - this can be physical, biological, social, psychological, conceptual or even another mathematical model".

Finally, it is necessary to make a conceptual and methodological approximation between modeling and the field of *applied mathematics,* the former of which is an indispensable tool for the latter. What's more,

---

[2]For this world-renowned Brazilian educator, education should serve to liberate and raise awareness among its subjects (AZEVEDO, 2010).

Mathematical construction can be understood in this context as an activity that seeks to synthesize ideas conceived from empirical situations that are almost always hidden in a tangle of variables. Doing mathematics, from this perspective, means combining abstraction and formalization in a balanced way, without losing sight of the original source of the process (BASSANEZI, 1999, p. 13).

The author also adds that mathematical modeling is what is known as the process of adapting a model to achieve certain objectives, adapting it to a better way, or analyzing it in a comparative way, taking an existing model as a reference. In this way, "the challenge for the teacher who adopts modeling as a teaching method is to help the student understand each stage of the process by building meaningful mathematical relationships" (BASSANEZI, 1999, p. 13).

1.2.6   Problem solving

There is a certain conceptual confusion in the interpretation of what a *problem is in* math teaching. This means that *problems do* not play the role of methodologies in the teaching-learning process and end up being used as tools in teachers' habitual practices (PEREIRA, 2008; BRASIL, 1998). Perhaps because of this conceptual proximity between methodology and tool, the imperceptibility of which degenerates any methodological proposal, David (1995) states that Problem Solving is the math teaching methodology that requires the least change compared to more traditional teaching.

According to Onuchic and Allevato (2004), the first systematic research on the subject began in the 1970s. The published *Curriculum and Evaluation Standards* reiterated that problem solving should be the main objective of all mathematics teaching and an integral part of all mathematical activity (ONUCHIC and ALLEVATO, 2004).

In this methodological approach, the importance of the correct answer gives way to the importance of the solving process, developed from a sequence of concatenated actions or operations, since what is precepted is that students learn a didactic path capable of not only indicating the correct answer, but also guaranteeing the appropriation of the knowledge involved (PEREIRA, 2008; BRASIL, 1998).

Teachers in the early years of elementary school can use Problem Solving to introduce concepts of basic operations and thus enable students to learn mathematics in the context of their everyday lives. According to Mendes (2006), this teaching methodology aims to develop cognitive skills, encouraging reflection and questioning, as opposed to memorizing and expository teaching. In the final grades of elementary school and in secondary school, the approach should be investigative, so as to encourage the exercise of raising and testing hypotheses in order to develop possible algorithms for solving the problem.

Thus, according to Flemming, Luz and Mello (2005, p. 74), "it is necessary to start from the simple to access the complex, and complex problems are visualized as a set of simple parts".

Problem-solving will make a greater contribution to learning when "the problems" relate to situations experienced by the students, until they move from the stage of solving problems to that of learning mathematics, i.e. learning mathematics to solve problems (MENDES 2006). By awakening the pleasure of solving problems, students can encourage themselves to propose other problems to their classmates and teachers, making them critical and challenging people who will not hide from problematic situations, but rather challenge them.

Also according to Mendes (2006), by using this technique, some students are able to transpose problems from their reality into deductive, inductive, spatial, graphical, proportional and other forms of reasoning that are abstract to their daily lives. They also manage to be more critical and constructive in mathematical situations.

Summarizing the methodological path that the teacher must follow and follow in applying Problem Solving, Onuchic and Allevato (2004) outline its phases starting with (i) the posing of a problem situation that expresses key aspects of the mathematical topic being dealt with, on which mathematical techniques must be developed in the search for at least reasonable answers, at first. In this stage of applying mathematical techniques, according to Flemming, Luz and Mello (2005), the first step should be to understand the problem, what has been provided (data) and what is being asked (unknowns). After this, the next step (ii) is to work out a solution strategy, checking the ways in which it can be solved, choosing the most practical and quickest way and carrying it out carefully. Finally, (iii) you must critique the results obtained, checking their contextualization.

These four steps were defined by Polya (1995) in his book "The Art of Problem Solving":

> In order to properly group the questions and suggestions on our list, we will distinguish four phases of work. Firstly, we have to understand the problem, we have to be clear about what is needed. Secondly, we have to see how the various items are interrelated, how the unknown is linked to the data, to get an idea of the solution, to

establish a plan. Thirdly, we execute our plan. Fourthly, we look back on the completed solution, review it and discuss it (POLYA, 1995, p. 3-4).

For Polya (1995), there is no point in solving something that has not been understood. For the author, once we have understood something, we have to understand the calculations, the drawings and the paths we have to follow, and then set about solving the problem, remembering that the work doesn't end with the problem solved. We need to look at the solution, draw conclusions and analyze the results. For this purpose, literal problems are more conclusive than numerical problems, because we can make conjectures and draw conclusions that are not possible with numerical expressions.

The technique of solving mathematical problems is mastered by few, and those who master it are able to visualize, understand and solve various problems in the most diverse areas. Good problem solvers have a logical sequence for their solutions, which starts with a good interpretation of the situation, extraction of essential data, use of algorithms and mathematical calculations, culminating in solutions rich in concepts and great interpretations. These people are able to transfer connections between mathematics and other concepts, both real and abstract, into their own lives (MENDES 2006).

It is important that during this problem-solving process, the problem situation is the starting point for the mathematical topic to be studied and not the definition of the topic itself, i.e. the knowledge to be produced and acquired is formed during all the stages. It is on the problem situation that the student applies non-mechanical knowledge after interpreting it, inferring first approximations of the result, which make sense through a process of rectifications and generalizations with the help of the teacher (BRASIL, 1998).

The activities proposed in this research work with Problem Solving, as we believe that this teaching-learning methodology is conducive to the development of concepts needed by students on the Agriculture and Husbandry course integrated into secondary education, as well as in all secondary education courses. It uses a methodology that involves the students' previous knowledge of mathematics and the contextualization of everyday farming situations, to introduce concepts useful to their profession. This activity also involves Information and Communication Technology, which we will discuss below, as a source of research and with the use of GeoGebra *software* to visualize geometric solids.

1.2.7 Information and Communication Technologies

The term Information and Communication Technologies (ICT)[3] has been used more prominently since the 1990s, when the main technological insertions began to appear in people's daily lives. While previously some ICTs were restricted and expensive, since then a variety of tools have become accessible.

Technological resources, increasingly emphasized by the media and exploited by the computer market, promise a revolution in schools and in teaching and learning. However, the educational trend that appropriates these resources, referred to here as *technologies,* should be viewed with caution within Mathematics Education, since using computerized resources does not necessarily mean that learning is enhanced. Computers, calculators and cell phones, among others, must be used in an orderly and coordinated manner by the teacher if they are to be put to good use.

Borba (1996), at the time his article was published, had already noted the significant changes that computers were bringing to the mathematics classroom, in terms of what should be taught and learned. According to him, these changes were not just about replacing one topic with another, but above all about the way the teacher would have to relate to the students and the machine.

It should be borne in mind that there is a common tendency, but obviously not the rule, for beginner teachers to reproduce in their daily practice the way their teachers taught them, especially those in their undergraduate studies. Borba (1996, p. 124) also reinforced the importance of the future teacher's contact with technologies during their initial training and foreshadowed a scenario that can be observed today, when he stated that if the points mentioned in the previous paragraph, among others,

> are not covered in teacher training, it is possible that we will have two scenarios when some schools have broad access to computers: the first is that teachers may only deal with various topics in the same way, simply by switching media. In this case, the computer is seen only as a "faster" notebook and/or book. The second scenario is that computers will not be used (BORBA, 1996, p. 124).

Furthermore, in addition to initial training, in order for teachers to be able to make use of these technologies, it is extremely important for them to undergo ongoing training, associated with working together with colleagues where they work or linking up with study groups, forming a web of information. With this, they will have the confidence to apply activities using computer resources. Penteado (2004, p. 287) states that:

> The use of computers in schools will not be consolidated with the support of sporadic

---

[3]In this study, as ICT will have a leading role, the term will be used, for the most part, simply as *technologies.*

courses for teachers from different locations and subject to different working conditions. At school level, teachers need to be motivated to organize and develop activities with computers and, in partnership with researchers, computer technicians, parents, students and other educators, they need to be able to create strategies to solve local problems.

This idea is in line with the concept presented by Miskulin et al. (2006), in which technologies represent the possibility of more cooperative learning experiences. According to the authors,

> the teaching proposal should be integrated with technology and collaborative methodological resources should be used to develop skills that the teacher will perform in the classroom, thus preparing the teacher to be the mediator who prioritizes technology in their workplace (MISKULIN et al., 2006, p. 108).

Frederico and Gianoto (2014) state that teachers need to plan their activities using computers, as well as have computer skills, so that unforeseen situations in their classes can be managed by the educator and do not compromise the content taught. For Penteado (2004), the application of these resources takes teachers into a risk zone where loss of control can occur at any time, and they don't feel comfortable with this situation. They therefore want to return to their comfort zone and, in this way, return to simply lecturing.

This question leads the study of mathematics to a good discussion. This debate, already raised in some studies, permeates the possibilities of using new media in the context of the classroom, in addition to the traditional pencil-and-paper.

For Borba (1996), mathematics has always been seen as an abstraction and therefore immune and not permeable to other media. Also, for Araújo et al. (2008, p. 11), "knowledge is constructed by human beings-with-media", where these media, considered in their plurality, include speech, pencil and paper, calculators, computers, among others, and, from this perspective, "the nature of mathematics constructed when computers are present is different from that constructed with human beings-with-pencil-and-paper".

It is well known that most schools in the country have access to technology through computer labs. However, according to Frederico and Gianoto (2014), the use of these spaces is being underutilized, because, for the most part, access to computers is restricted to using the internet for research, and there is little use of *software and* spreadsheets, where there is greater application to the teaching of mathematics.

Computers offer a number of spreadsheets that work on mathematical concepts such as functions, statistics, matrices, determinants and others that can be explored by teachers in their classes. In addition, there is *software* that works with graphs, geometry, trigonometry, among others, which involve many abstract concepts for students and which have tools that can make these concepts more accessible. As Frederico and Gianoto (2014, p. 68) quote:

> Computers and the internet have caused major changes in the daily lives of children, young people and adults. The internet, *software* and games, among others, have gained prominence as tools for entertainment, research and schoolwork. Using text editors, for example, you can carry out numerous activities, such as typing, editing images, inserting pictures, tables, etc. With spreadsheets, you can perform a series of calculations and assemble graphs and tables, among other activities. Of course, you can't forget the virtual encyclopedias and the variety of *websites* that serve as a source of research.

Computer resources will always be available to be applied in the teaching-learning process, but they shouldn't just be used as didactic enrichment, they also need to be part of a knowledge process that will be of extreme personal use to the students, as they will be inserted into the job market, where computer skills are a prerequisite for jobs that require more qualifications and almost always offer more attractive salaries.

In this sense, it can be seen that

> many Brazilian schools have not fulfilled the task of preparing students for the technological world, which is no longer an intellectual abstraction, but a reality that is being imposed more and more intensely, and which must be faced by reflecting on and reshaping the ways of teaching mathematics, adapting them to the demands of a computerized society (MISKULIN et al., 2006).

From the perspective of using some media that involve technology, mathematics should not be mediated by obsolete models that contribute little or nothing to the development and transformation of the individual in training, but by alternative methodologies in which new educational processes are implemented that make sense and are related to their integration into society (MISKULIN et al., 2006).

In addition, Miskulin (1999) discusses the need to introduce and use computers with students from different social classes in an attempt to reduce inequalities:

> Technology is not just another resource for teachers to motivate their classes, it is a powerful medium that can provide students with new ways of generating and disseminating knowledge. Therefore, mathematics teachers should think about [...] creating projects in schools that can provide opportunities for students to learn mathematics and, at the same time, use technology in such a way that mathematics, in a technological context, becomes a pathway that can overcome social inequalities and also make it possible to train people adequately for the job market (MISKULIN, 1999, p. 4).

He goes on to say that efforts should be made to create learning environments with technological resources available to students and, above all, with an up-to-date pedagogical proposal that takes into account advances in technology (MISKULIN et al., 2006).

From a perspective that could be considered audacious, the author is still moving towards a scenario where technologies applied to education must migrate from laboratories separate from the classroom to a concept that integrates them with the development of themes related to the various areas of knowledge. Thus,

> technology becomes a tool that can be accessed within the classroom itself, becoming a pedagogical resource to support the teacher in developing the lesson plan. This perspective allows students and teachers to become immersed in the topic under discussion, stimulating and creating new skills for the development of logical, communicative and creative reasoning (MISKULIN et al., 2006).

Thus, in this chapter we present some of the trends in the teaching and learning of mathematics, as well as a brief introduction to mathematics education, since the topics covered in this work are problem solving and technologies.

In the next chapter, we will describe the evolution of the teaching-learning methodology of Problem Solving and the use of Technologies in education, as well as the justification for using the two teaching trends in one activity, to be used in the classroom in activities for learning Spatial Metric Geometry.

CHAPTER 2

Problem solving combined with IT: A proposal for use in the classroom

This work focuses on what Mathematics Education can contribute to students' learning, so we address the trends of Problem Solving and Technology, considering the importance of these two trends today. The former has been used for a long time, since the end of the last century. The second, on the other hand, is much newer, having emerged in the 1990s with the popularization of computer resources, and is currently dispensable to human needs. By combining these two trends, we believe we can achieve educational success in the teaching of geometry, which is feared by students for its lack of spatial awareness and difficulty in relating the plane and the sword.

Mathematics Education, based on its trends, seeks to bring mathematics closer to the people, because it sees the need for mathematical concepts in their daily lives. But we can't conclude that the alternatives offered by Mathematics Education are the only solution to solve the problems of teaching mathematics. Onuchic (2013) comments that Pure Mathematics should not be discarded, at times it is extremely necessary, because we have a predefined curriculum and we cannot fail to comply with it.

We know that Mathematics Education, despite providing benefits, has some barriers and the main one is the time spent on its applications. We have to remember that students don't only study Mathematics, we can't work all the time with trends in Mathematics Education. From another angle, the majority of students do not have a good grounding in the concepts of pure mathematics. Onuchic (2013, p.91) also reports this parallel and shows the need to unite these two ways of working with mathematics. On the one hand, he cites an article by Bass, which shows the importance of mathematical concepts to provide the language and concepts suitable for description, analysis, modeling and simulation, and on the other hand, Mathematics Education which:

> leads to intense debates among mathematics teachers at all levels of education, mathematics educators moving into a field of study, mathematicians collaborating on curricula, with their concepts and contents, their operational techniques and their many different applications. (ONUCHIC, 2013, p. 91)

2.1 Approaches to Problem Solving

Although history shows that problem solving has been used since the time before Christ and was considered a teaching trend at the end of the last century, there are disagreements about how to apply this technique to the teaching and learning of mathematics. When, in 1980, the *National Council of Teachers of Mathematics* (NCTM) in the United States published the document *An Agenda for Action: Recommendations for School Mathematics of the* 1980s, concerned about the teaching and learning of mathematics, it defined as its first recommendation that problem-solving should be the focus of school mathematics in the 1980s (NUNES, 2010). According to Nunes (2010, p.80), "Mathematics educators at that time were very interested in making problem solving a focus of the mathematics curriculum". However, without a well-defined methodology, there were several different approaches and researchers began to question this methodology, making it the focus of states in the 1990s.

Problem solving methodology should not be confused with solving problems in order to establish mathematical concepts. Brasil *apud* Allevato (2005) states:

> Traditionally, teachers use problems to check and fix learning. However, if we look at the history of science, we can see that the problem invariably precedes the discoveries, it is the provocateur of studies and the guide of theoretical constructions. Why do we invert the natural order of things when teaching mathematics in particular? (BRASIL *apud* ALLEVATO, 2005, p.22)

Problem-solving needs to be approached in such a way as to build new knowledge and not to recall or fix knowledge that has already been acquired; from this point of view, we should use it as a driver of learning. According to Damaceno (2011, p. 6):

> We can use problem solving as a starting point for introducing content. Often, some subjects are overloaded with information, concepts, definitions, in short, it is inevitable that in many cases some students' learning will be blocked. This happens because they are unable to associate this new knowledge with their own baggage of information. However, with a good choice of problem, it is possible to build a bridge with the knowledge that the student has already acquired, promoting greater

understanding for the introduction of the desired new content. (DAMACENO 2011, P-6)

A problem that is well designed by the teacher should contain concepts that the students are familiar with and that are close to their realities, so that it can be interesting to solve. With this awakening, the students may want to develop the problem and, at that point, they will find concepts that they need to research in order to learn them. In this way, the teacher may be able to introduce concepts with Problem Solving and not use it as a learning analyzer.

With the interpretations of "An Agenda for Action" in the 1980s and the studies carried out in the following years, problem-solving, according to Nunes (2010), received four different approaches to teaching mathematics: teaching *"about"*, *"for"*, *"via"* and *"through"* problem-solving, although prior to Nunes, some authors did not differentiate between the last two approaches. Let's talk a bit about each of them.

2.1.1   Teaching Problem Solving

It is a method in which the teacher teaches students how to solve a problem, and it is a new concept, a methodology. Nunes (2010) says that educators who teach problem solving follow Polya's four steps or some variation of them, which are: "understanding the problem; devising a plan; taking the plan forward; and looking back at the original problem in order to analyze the validity of the solution found" (NUNES, 2010, p.82). In this methodology, students learn to be good problem solvers, but mathematics is not always learned using this technique.

2.1.2   Teaching *for* Problem Solving

Teaching *to* solve problems addresses learning mathematics for use in solving problems, both closed and open. In order to verify the approaches developed for problem solving, we must define the types of problems, which according to Shimada *apud* Allevato (2005, p. 43-44), are twofold: closed and open. The first is made up of traditional problems, in which there is only one correct and predetermined answer, while the second has various methods for obtaining the answer.

The author states that the latter should be the first to be introduced to students, so that they can experience new things in this teaching-learning process.

This methodology is used in a repetitive way, in which students do a large number of problems, using the material worked on in class, using some as models to solve others, as a way of consolidating the content covered. According to Nunes (2010), "students should be given many examples of the mathematical concepts and structures they are studying, and many opportunities to apply this mathematics to solve problems". The author claims that this approach is justified by the concept that the purpose of learning mathematics is to be able to solve problems. From this point of view, teaching *to* solve problems shows the application of the concepts worked on and the need for students to learn them in order to apply them in their daily lives.

2.1.3   Teaching *through* Problem Solving

At the end of the 1980s, scholars warned against misinterpreting the main recommendation of the "Agenda for Action", which was that problem solving should be the focus of mathematics in the 80s. They began to think of problem solving as a methodology for teaching mathematics, as a driving force for knowledge and no longer as a method for fixing concepts (NUNES, 2010).

With these studies, they concluded that teaching *through* problem solving was the way to work with mathematics. As Schroeder and Lester *apud* Nunes (2010) state

> In problem-solving teaching, problems are worked on not only for the purpose of learning mathematics, but also as the main means of doing so. In this approach, the teaching of a math topic begins with a problem situation that incorporates key aspects of the topic, and mathematical techniques are developed as reasonable answers to reasonable problems. One aim of learning mathematics is to turn certain non-routine problems into routine ones. Mathematical learning, in this way, can be seen as a movement from the concrete (a real-world problem that serves as an example of a mathematical concept or mathematical technique) to the abstract (a symbolic representation of a class of problems and techniques for operating with these symbols). (SCHROEDER AND LESTER *apud* NUNES 2010, p.84)

Allevato (2005) comments that students are no longer seen as empty vessels to be filled with fragmented information, but as part of the teaching-learning process, according to a constructivist idea.

Nunes(2010) and Allevato(2005) comment that teaching through problem solving is the approach most consistent with the recommendations of the NCTM[4] , according to which:

National *Council* of Teachers *of* Mathematics

(1) Mathematical skills and concepts should be learned in the context of problem solving,

(2) the development of higher-order thinking processes should be stimulated through problem-solving experiences, and

(3) the teaching of mathematics should take place in a problem-solving environment.

2.1.4   Teaching *through* Problem Solving

Teaching *through* problem solving is just a conceptual variation of teaching *through* problem solving, because the *via* approach means "through", while the expression *through* is more comprehensive, it accompanies learning from start to finish and is not just a teaching-learning resource. Nunes (2010) comments that:

> The expression "through" is a way of teaching and, consequently, learning and, in the process, doing mathematics, since the student, faced with a problem, must show himself to be a co-constructor of his own knowledge. In this approach, the primary objective is to present students with problems that will generate new concepts or content (NUNES, 2010, p. 84 and 85).

Van de Walle *a/pud* Nunes (2010) emphasizes that the teaching of mathematics should start from the students' knowledge and not from the teacher's knowledge, as in traditional methods. The students' knowledge must be taken into account so that they can build their own knowledge, rather than having gaps filled in by information from teachers, without paying attention to the different ways in which each person interprets and carries out everyday activities.

The classroom environment needs to contribute to this. Teachers need to think about and be well prepared for three important moments: *before, during* and *after*. With well-planned actions by educators, following Van de Walle's understanding, Nunes (2010) expresses the three moments with the following approaches:

- *before:* the concepts previously acquired by the students should be emphasized, in addition to clearly explaining the objective of the proposed problem, so that at the end of this moment there are no doubts about the lesson proposal.

- *during:* dividing the class into groups, which can approach the problem in different ways. In this case, the teacher needs to know how to listen, without directing them to solve the problem, so that different ways of solving the problem emerge.

- *afterwards: at* this point, the teacher needs to encourage discussion of the solutions obtained by the groups, without evaluating them, leaving it up to the students to decide on the best way to solve the problem. After this socialization, the teacher should conclude the formal way of solving the problem and not discard the other solutions, encouraging them to reflect on the methods and extensions of these solutions.

The problem-solving methodology tries to bring the learning of mathematics closer to a pleasurable situation, because everyone can participate without fear of "making mistakes", because they are in the process of learning, in which mistakes are constructive for teaching and learning. Allevato (2005) points out that:

> When the teacher adopts this methodology, students can learn both *about* problem solving and learn mathematics *to* solve new problems, while learning mathematics *through* problem solving (ALLEVATO, 2005, p. 61).

The "Problem Solving" trend in mathematics education has gone through all the stages of *about, for, via,* and is currently only considered *via* as a teaching approach.

2.2   IT

Among the various technological resources available, information technology has been innovating teaching and learning, which is somewhat underused by teachers, as mentioned above, because it takes the teacher into a risk zone, where he or she does not have total control of the proposed activities (BICUDO and BORBA, 2004). With better prepared teachers and the use of didactic *software*, we can improve the teaching of mathematics, because the *software* is dynamic, engaging and visualizes the interaction of algebraic, geometric, statistical and trigonometric concepts, graphs and tables.

There is *software* with a paid license, *Shareware* (can be tried before buying), demo (demonstrates what it does and can be manipulated with some features) and free (with free installation). This work will use the free GeoGebra Dynamic Geometry *software,* as it is easy to install and manipulate. According to Araújo (2010),

> GeoGebra (Geometry and Algebra) is an open-source program (GNU - General Public License), which can be downloaded for free from www.geogebra.org. Some facts available from the software's website (GEOGE-

BRA, s/d) include the fact that GeoGebra is "a free, multi-platform dynamic mathematics software for all levels of education that combines geometry, algebra, tables, graphs, statistics and calculus in a single system". (ARAÚJO, 2010, p. 49)

Using free pedagogical *software* allows students to install and manipulate it at home, making the learning activity transcend the school. The students involved in the technology are encouraged to produce other didactic activities that will help them understand the geometry they work on in high school.

2.3   Problem solving combined with IT

Problem-solving combined with IT is the union of two teaching-learning trends, the aim of which is to use IT as an instrument for visualizing and experimenting with the learning acquired through problem-solving, as well as producing new knowledge. From this perspective, Borba and Penteado *a0ud* Allevato (2005) reject the dichotomous view between human beings and technology, stating that:

> human beings are made up of techniques that extend and modify their reasoning and, at the same time, these same human beings are constantly transforming these techniques. A dichotomous view therefore makes no sense. What's more, we understand that knowledge is only produced with a particular media, or with an information technology. This is why we have adopted a theoretical perspective that is based on the notion that knowledge is produced by a collective made up of human beings-with-media, or human beings-with-technologies and not, as other theories suggest, by solitary human beings or collectives made up only of human beings. (ALLEVATO, 2005 p. 46)

Problem-solving needs to be approached in such a way as to build new knowledge and, for this to happen, the classroom environment needs to contribute to this teaching, Allevato (2005) comments that:

> Observations made by researchers show that, during teaching situations in a computer environment, students generally choose to work in groups and tend to discuss mathematical activities with more interest. Since computers began to be introduced into teaching around 1980, there have been reports of a marked trend towards the development of collaborative learning environments. Based on *feedback* provided by the computer, students begin to exchange experiences, share understandings, give suggestions to their classmates and go through a game of counterexamples, new conjectures and reformulation of concepts (ALLEVATO, 2005, p. 90).

When we look for a comparison between the two teaching trends, we can see that both seek to teach by a similar sequence, because for Valente (1993), the sequence used in Technology is: the description of an idea in a formal language; then the execution and description of the data for a result obtained by the machine; in the third step, the student will have to reflect on these results; and finally, a possible debugging, if the expected result was not obtained. These four acts cited by Valente (1993) are similar to Polya's four steps in Problem Solving, which are: "understanding the problem; devising a plan; carrying out the plan; and looking back at the original problem in order to analyze the validity of the solution found" (NUNES, 2010, p.82).

In this perspective, these tendencies interact, because describing an idea is similar to understanding the problem; executing and describing the data is equivalent to analyzing a plan; carrying out the plan and analyzing the validation of the solution are analogous to reflecting on the results and a possible debugging.

It is important to emphasize that there is a need for ongoing training for the teacher who mediates this process, since the teacher must have mastery of the *software* used in order to be successful in the course of the work.

This work will use printed, manipulative and digital technologies, especially the GeoGebra *software*, to teach through Problem Solving, with an approach aimed at students on the technical course in agriculture and livestock farming. With the combination of these teaching methodologies, we believe that the result will be more satisfactory, as the technology is used by all the students and Problem Solving involves situations from their daily lives. According to Gomes and Rodrigues (2014), "the teacher ends up using many tendencies in a given activity", so sometimes this combination is due to their training and sometimes because they use their creativity in an attempt to improve students' teaching and learning.

In this sense, in the next chapter, we will present a proposal that involves combining these trends to introduce the concepts of Spatial Metric Geometry.

CHAPTER 3

Calculating the volume and area of geometric solids through Problem Solving and Technology

According to Nunes (2010), the teaching of mathematics in Brazil in the last century was fragmented into Arithmetic, Algebra and Geometry, subjects taught by different teachers in most schools. According to Nunes (2010), this meant that the subjects were taught in an abstract way, without contextualization. With the Modern Mathematics Movement, around the 1960s, there was a greater preoccupation with algebra and arithmetic, with a relative disregard for geometry, as it has a more questionable form of its concepts, from students to teachers. Nunes reports that Geometry was not taught in many public schools, which led to a lack of knowledge among students and learning difficulties for future teachers. At the end of the 1980s, there was a movement of teachers who were concerned about the lack of knowledge of Euclidean concepts of geometry, and promoted an effort to resume the teaching of this content in all schools. Until, at the beginning of this century, Geometry came to be seen as a driver of other knowledge, as Nunes (2010, p. 101) mentions:

> The PCNs[5] (2001), in turn, emphasize the importance of teaching Geometry in school curricula, when they justify its relevance in terms of work where geometric notions contribute to learning numbers and measures, as it encourages children to observe, perceive patterns and identify regularities and vice versa. (NUNES, 2010, p. 101)

Geometry should be taught in primary and secondary schools with an emphasis on forming critical students capable of better logical reasoning. For Nunes (2010), based on Standards 2000, teaching geometry efficiently is justified by:

> (1) when studying geometry, students have the opportunity to learn about geometric shapes and structures and how to analyze their characteristics and relationships; (2) spatial visualization is an essential aspect of geometric reasoning; (3) Geometry is a natural context for developing students' reasoning and argumentative skills, culminating in demonstrated work in secondary school; (4) Geometric ideas are very useful in representing and solving problems in other areas of mathematics and in everyday situations, so Geometry should be integrated with other areas whenever possible. (NUNES, 2010, p. 107)

In order to develop knowledge of geometry, students need to be motivated and since this motivation is not achieved externally, we need to encourage them to use their knowledge and skills to solve problems in their daily lives. That's why we've combined Problem Solving and Technology to solve a Spatial Metric Geometry problem, which is used in the training of students on the Agricultural Technician course integrated into secondary school.

Spatial Metric Geometry studies geometric solids, their measurements, their elements, and the calculation of their areas and volumes. These solids are divided into two classes, polygonal and round, the former being prisms and pyramids, the latter being cylinders, cones and spheres. The aim of this work was to develop these learning concepts using Problem Solving combined with Technology, as we proposed a teaching activity using Problem Solving, two secondary problems[6] and an activity using GeoGebra. With this, the students can use the technologies in order to be instructed, grounded and prepared to solve the problem of learning these solids.

3.1 Choice of mathematical problem

When thinking about a proposal to be used in the classroom, we are faced with the following question: what is the difference between a math problem and an exercise?

For Allevato (2005), an exercise, in the form of a problem, is for re-remembering, practicing, exercising concepts already acquired in solving an algorithm. A mathematical problem needs to be investigative; you can't have all the pre-defined ways of solving it. According to Allevato (2005) and Onuchic (2004), a question is a problem if the student does not yet know the steps to solve it, but is interested in solving

---

[6]This will be discussed in more detail in section 3.1.1

it.

To make this choice, we must use what the students have already learned, so that they can use it as a tool, even if we need to remind them, and it also needs to be part of their daily lives, so that they can see how useful it is in their lives. The problem can't be too easy, where the students already have all the learning paths, but it can't be too difficult, where they don't have a starting point or an understanding of what they need to do. The problem should be challenging, so that students can search for new concepts based on definitions they have already acquired.

In order to teach geometric solids, which is a subject in the second year of secondary school, defined by the Common Basic Content of Mathematics in the state of Minas Gerais, we will propose a problem of calculating area and volume, which is a new concept, using content already seen by students in this grade. This activity will be linked to agriculture and will use concepts from Plane Geometry, such as area, the Pythagorean theorem and the circular sector, as well as concepts from the technical area.

We will use secondary problems to remind students of the definitions and applications of Plane Geometry. We will mention agricultural concepts in the problem itself, as this is not the focus of our work, even though it is relevant for motivating students.

3.1.1 Secondary problems

Even with the secondary problems, students are still challenged and continue to work with Problem Solving, rather than proposing a review, which is often what is practiced in class.

For Nunes (2010), a secondary problem can be:

> Doubts presented by the students in the context of the vocabulary present in the statement; in the context of reading and interpreting: in addition to those that may arise when solving the problem: notation. transition from vernacular language to mathematical language, related concepts, operational techniques, in order to enable the work to continue. (NUNES, 2010, p. 92)

To help with the problem-solving activity, we can use some smaller problems that will help us during the activity. According to Onuchic (2004), these problems are recommended so that the teacher doesn't interfere by explaining how to solve the problem, but rather provides another "smaller" problem with concepts already known by the students.

We will propose two problems that recall the concepts of area, the Pythagorean theorem and the elements of the circular sector. We will use agricultural concepts to approach the secondary problems, so that students don't lose interest. These activities can be proposed in the course of the main problem, as soon as any doubts arise in a group of students, and are unnecessary for groups that manage to solve the problem without any doubts. Other doubts may arise and the teacher should be attentive to proposing other secondary activities.

3.2 The Problem Solving Activity

Computers, as well as printed bibliographies, will help as a source of consultation and useful information for the teaching and learning of solids. The construction of the solids in cardboard, EVA, acrylic or other materials, provided by the teacher or purchased by the students, will also be very important for the students to be able to visualize the solids based on their planning. It's important to always start with paper or cardboard so that more expensive materials aren't wasted and, with this attitude, you can insert an idea of a prototype, with low-cost materials, for conscious consumption, or, in the case of prisms and pyramids, the use of GeoGebra.

The aim of this work is to develop an example of how to work with problem solving in the classroom. We will apply the concepts of teaching through problem solving with an activity that uses the previous concepts of students on the technical course in Agriculture, in the vocational area and in secondary school, to teach and learn geometric solids (areas, plannings and volumes).

We believe that, for a better understanding, we will present the secondary problems before the main problem. However, as already mentioned, these problems can be proposed during the main problem.

3.2.1 The first activity: area and the Pythagorean Theorem

(1) Aim: to remember how to calculate the area of some figures and the Pythagorean theorem. In this activity, we will look at the area of an equilateral triangle, a square and a circle, remembering that we can use the area of a square to remember the area of a rectangle, and the Pythagorean theorem;

(2) Rationale: In the main activity, we will use these concepts to calculate the areas and the lateral apothem of the pyramid and the generatrix of the cone;

(3) Description of the problem: a small farmer has a horse and wants to fence off an area for its exclusive use. He found that he had enough material to build 360 meters of fence and, with his knowledge of geometry, decided that the shape should, when polygonal, be regular. What choice should the cattle farmer make,

knowing that he intends to use one of the three geometric figures, the triangle, the square or the circle, for his decision to consider the one that provides the greatest area for the horse? What are the dimensions (sides or radius) of this pasture?

(4) Skills: calculating the area of polygons and determining the elements of the right triangle, as well as the Pythagorean relationship;

(5) Technology: notebook, pencil, eraser, book, simple research using a cell phone or computer;

(6) Script: a possible solution script would be to determine the sides of the triangle and square, and the radius of the circle; from the side of the triangle, use the Pythagorean Theorem to determine its height; with these elements, calculate the areas of the figures and check which is the largest area;

(7) Evaluation: check, by means of other questions, whether the student knows how to calculate the area of triangles, quadrilaterals and circles, as well as their interest in carrying out the activity;

(8) Possible solution: for the triangle, the student will find a side of 120 meters, with a height of 104 meters, giving an area (S) of approximately 6,235m^2 , to arrive at this area, he will probably use the Pythagorean theorem to determine the height. In the square, the side will be 90 meters and it will have an area of 8,100m^2 . In the circle, the radius will be 57.30 meters with an area of approximately 10,315m^2 .

Triangle:

$$3 * l = 360 \Rightarrow l = 120m$$

$$l^2 = (\frac{l}{2})^2 + h^2$$

$$120^2 = 60^2 + h^2$$

$$h^2 = 10800 \Rightarrow h = 103,92m$$

$$S = \frac{b * h}{2}$$

$$S = \frac{120 * 103,92}{2} \Rightarrow S = 6.235,20m^2$$

Square:

$$4 * l = 360 \Rightarrow l = 90m$$

$$S = l^2 \Rightarrow S = 90^2$$

$$S = 8.100m^2$$

Circle:

$$2 * \pi * R = 360 \Rightarrow R = 57,30m$$

$$S = \pi * R^2 \Rightarrow S = \pi * 57,30^2$$

$$S = 10.314,76m^2$$

Figure 3.1: Illustration of the answer to the first activity

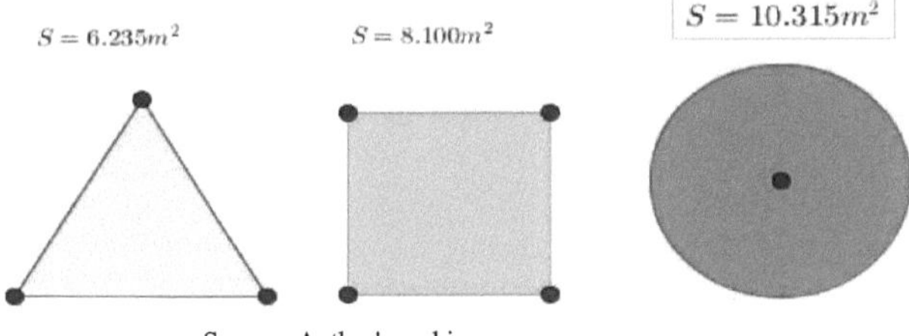

Source: Author's archive

3.2.2   The second activity: circular sector

(1) Aim: to cover the circular sector, in which we will calculate the elements of the sector (angle and arc length) as well as its area;

(2) Rationale: In the main activity, we will use these concepts to calculate the lateral area of the cone and

the ratio of the length of the base to the angle of planing and the generatrix of this cone;
(3) Problem statement: In a corner of a pasture, where the two fence segments form 60°, a ram is tied with a 30-meter rope from the ram to the corner of the pasture. In order to free the animal from the rope, its owner decided to build a fence in the area bounded by the rope. Based on this information, determine:
    (a) a geometric drawing of the situation described by the problem;
    (b) the geometric name of this design;
    (c) the area delimited by the rope for the sheep to graze;
    (d) the length of the fence that the owner will have to make to free the animal from the ropes.
(4) Skills: calculating the area of circular figures;
(5) Technology: notebook, pencil, eraser, book, simple research using a cell phone or computer;
(6) Script: a possible solution script would be to visualize that the ram describes a circular sector; transform the angle from degrees to radians; use a rule of three to determine the area of the circular sector; relate the length of the circular sector to its radius and central angle, to determine the length of the fence;
(7) Evaluation: check, through other questions, whether the student knows how to calculate the area, length and central angle of the circular sector, as well as their interest in developing the activity;
(8) Possible solution: the above activity will lead students to remember the circular sector; the area of the sector; the arc length of the circle; the relationship between central angle, radius and arc length. The drawing described by the problem is of a circular sector with a central angle of 60°. The area bounded by the rope is approximately 470m² and the fence to be built is approximately 31 meters.
Area:

$$S = \frac{\alpha * \pi * R^2}{360°} \Rightarrow S = \frac{60° * \pi * 30^2}{360°}$$

$$S = 470m^2$$

Rope length (1):

$$S = \frac{l * R}{2}$$

$$470 = \frac{l * 30}{2} \Rightarrow l = 31m$$

Figure 3.2: Illustration of the answer to the second activity

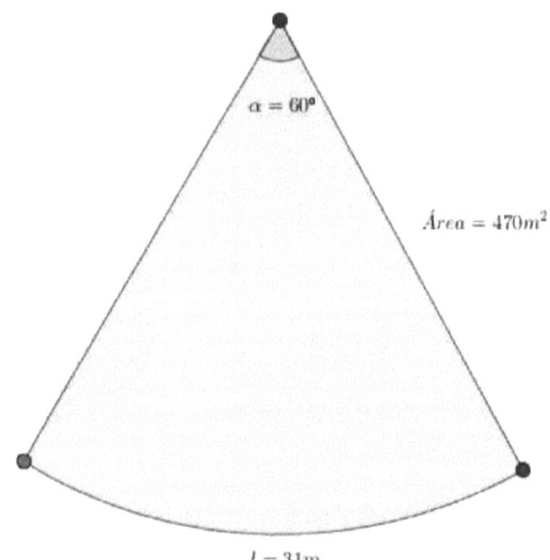

Font-c: Author's archive

The way to assess problem-solving activities is through interest in carrying out the proposed activities, since Onuchic (2004) comments that assessment should be based on student involvement and not on successes during the activity. Once the activity has been completed, a knowledge assessment can be carried out, with less complex questions and more focused on what you want to check.

As the purpose of this work is to combine Problem Solving and Technologies, we will use mathematical *software* so that the alliance is not just the use of Technologies as a tool for researching and recording learning, but as a tool for teaching and learning Mathematics.

Computers are a part of most students' lives and they are fascinated by the ease generated by this technology, which is why we are going to show them GeoGebra. This *software combines* technology with geometry (plane, spatial and analytical) and can also be used to study fungi. We will show some applications for solids, so that the students can use them to solve the problem. This third activity will last 90 minutes.

3.2.3   Third activity: use of technologies

This activity will focus on using the GeoGebra *software*, so that students can take advantage of this technological resource and get to know the prism and, with this activity and the teacher's encouragement, explore the other solids.

(1) Objective: to work on Spatial Metric Geometry using GeoGebra *software;*
(2) Reason: to get to know GeoGebra's tools and become familiar with the *software in* order to perform some of the calculations requested;
(3) Description: In a computer lab, with GeoGebra previously installed, suggest that students, individually or in pairs, access the program and, with a projector, show that GeoGebra has several windows. We will use the algebra window, the visualization window and the *3D visualization window,* where we will be able to see the algebraic elements, the plan and the solid, respectively, with the last window having to be enabled in the "View" and "3D visualization window" options, as can be seen in figures 3.3 and 3.4. For a better view, reduce the central window and move the X and Y axes;
(4) Skills: building geometric solids, visualizing their plannings;
(5) Technology: computer with GeoGebra *software* installed;
(6) Script: will be presented during the activity;
(7) Evaluation: to see if the student has learned to work with and use the *software, through* other similar activities, such as building another solid among those that will be covered in this work, as well as through the students' participation in the construction of the prism.

Figure 3.3: Display 3D window

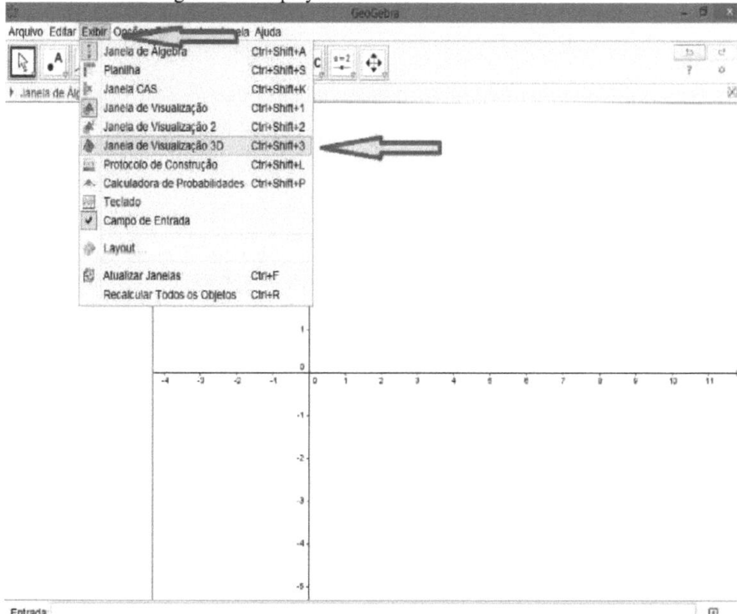

Figure 3.4: Show 3D window

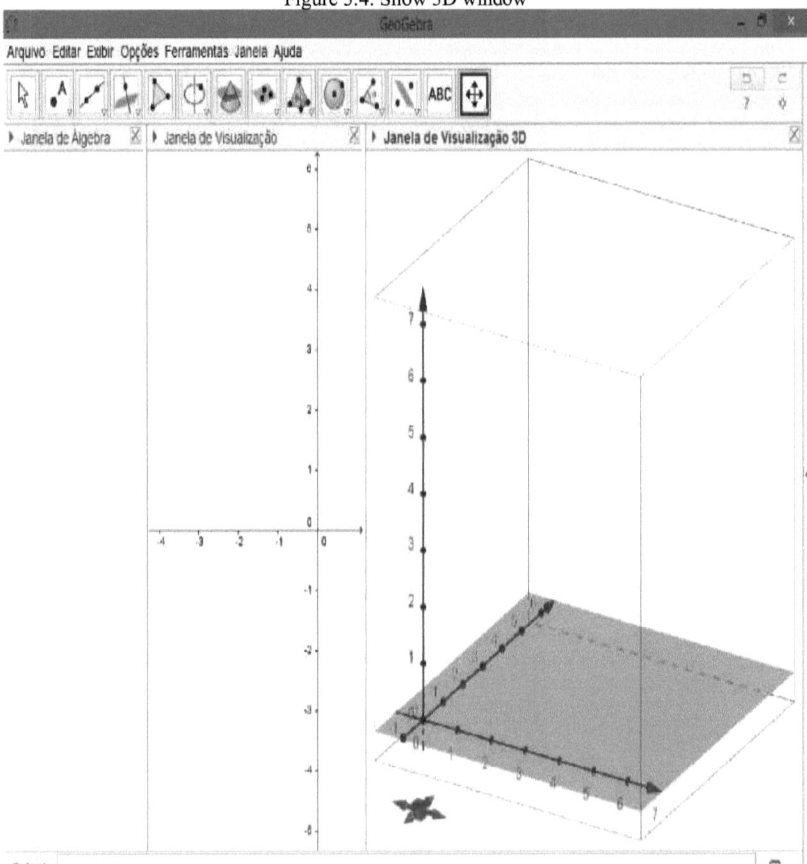

Students should follow the steps described on the following pages.

Select the (central) display window (1) and note that at the top there are some quick command windows, select the "slider" (2), second from right to left, and click on the top left corner of the display window, a window will open in which you can choose: the name, minimum value, maximum value and increment. Check "integer" (3) and choose the letter "a" for the name (4) and the following will appear in the window: Min "1" (5) (minimum value), Max "30" (6) (maximum value) and increment "1" (7), then click "OK" (8). The "a" element will give the size of the base side of the prism. These steps are illustrated in figure 3.5

Figure 3.5: Insert "a" slider

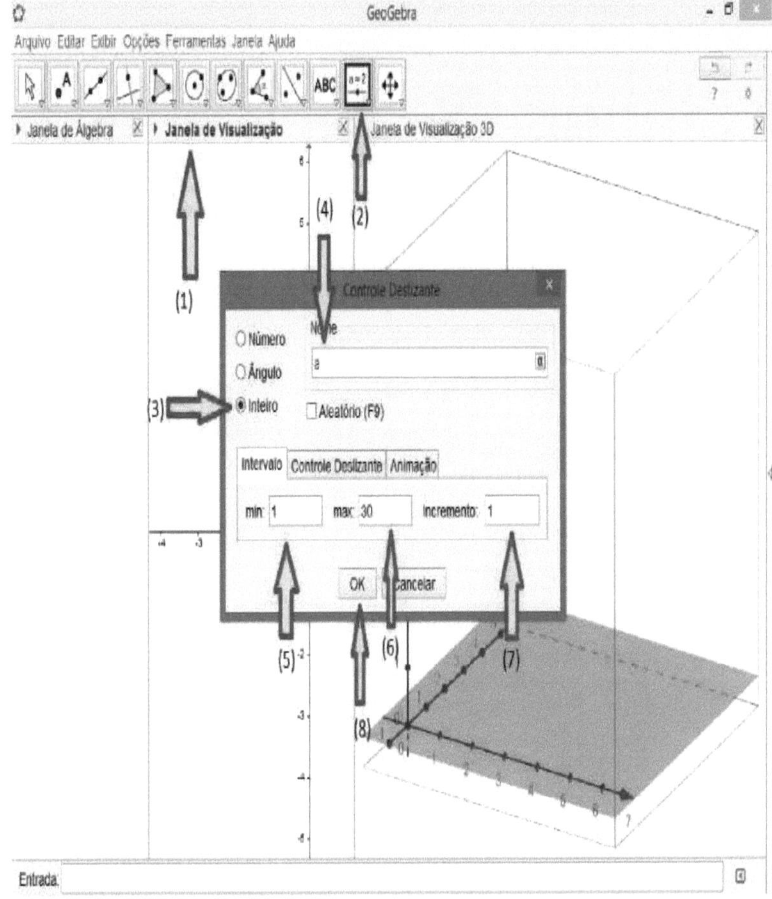

Source: Author's archive

Repeat this process for the other sliders, such as the letters "n" (number of sides of the base polygon), "h" (height of the prism) and "b" (will define the prism as straight or oblique), positioning them one below the other, only for the letter "n", change the "Min" value to " 3", as the "n" will give the number of sides of the polygon.

Figure 3.6: Inserting the other sliders

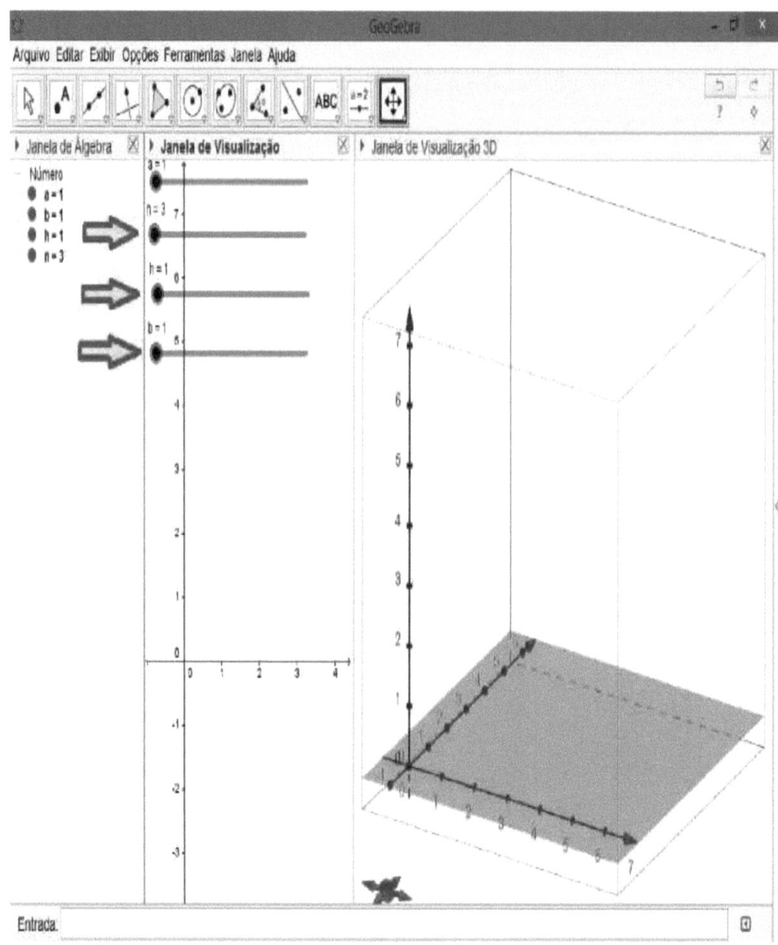

Source: Author's archive

In the "Input" bar, located at the bottom and on the left-hand side, type "A= (1,1,0)", then press *"Enter",* this will locate a point "A" in the XOY plane.

Figure 3.7: Locating point A

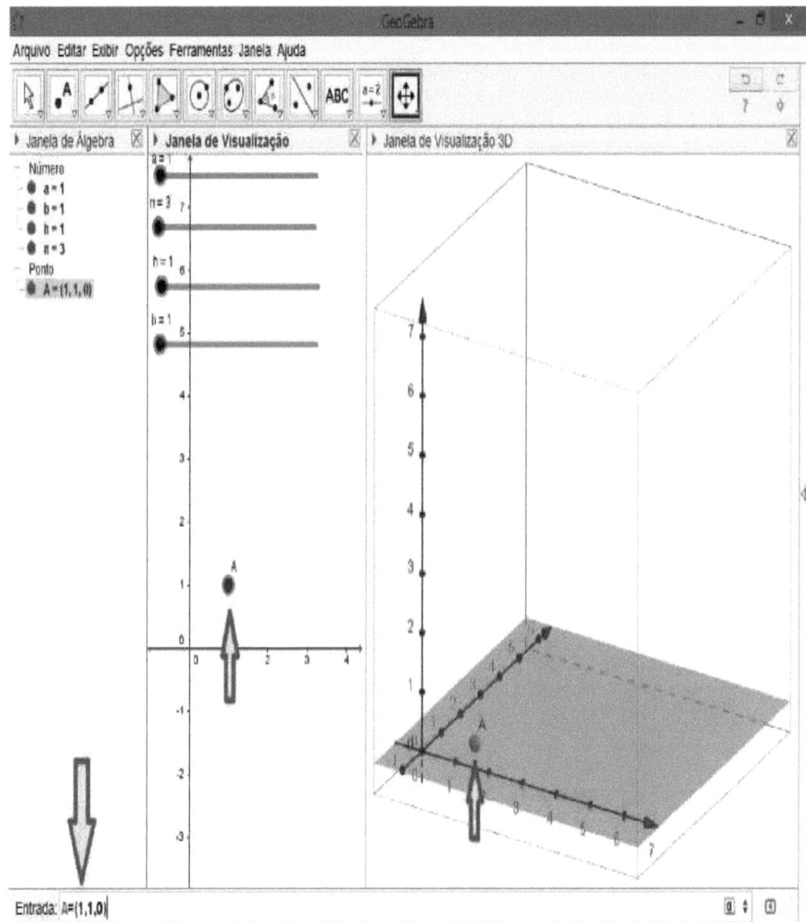

Source: Author's archive

In the same place, type "B=(1-a,1,0)" and type "*Enter*", a point "B" will be located on the same plane, at a distance of "a" units from point "A".

Figure 3.8: Locating point B

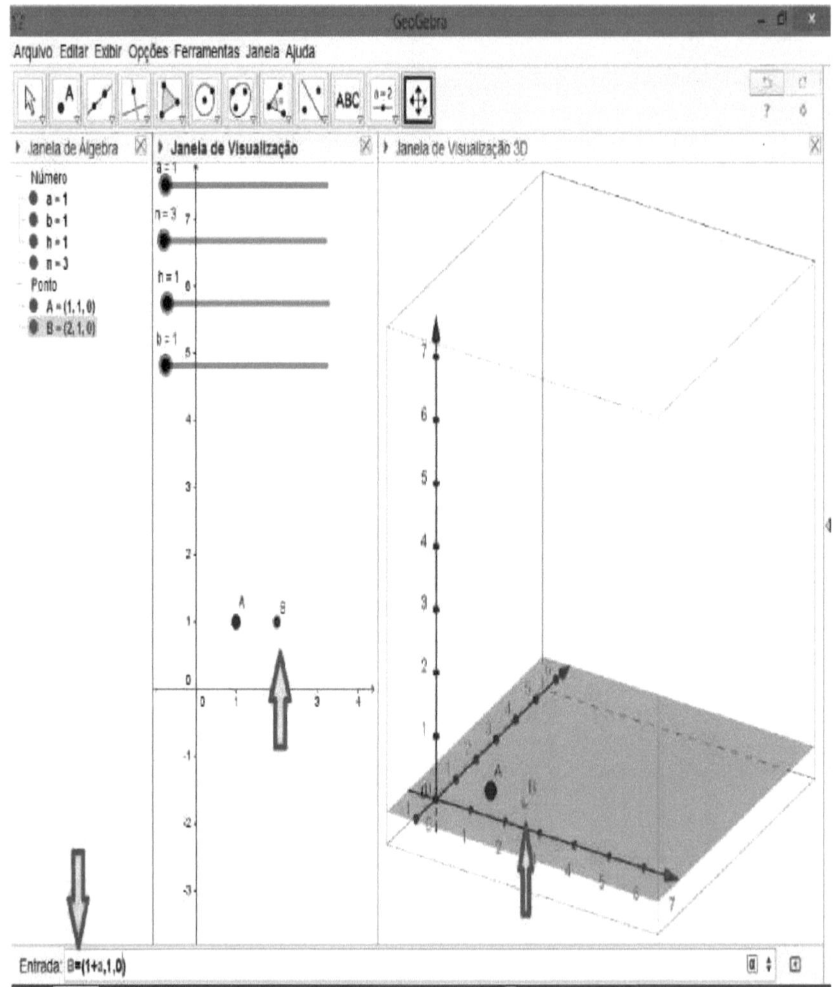

Source: Author's archive

In the "Input" box, type "Polygon[A, B, n]" and press " *Enter\ a* regular polygon with "n" sides, each with a size of "a", will be drawn in the visualization window and in the *3D* visualization window.

Figure 3.9: Construction of the base polygon

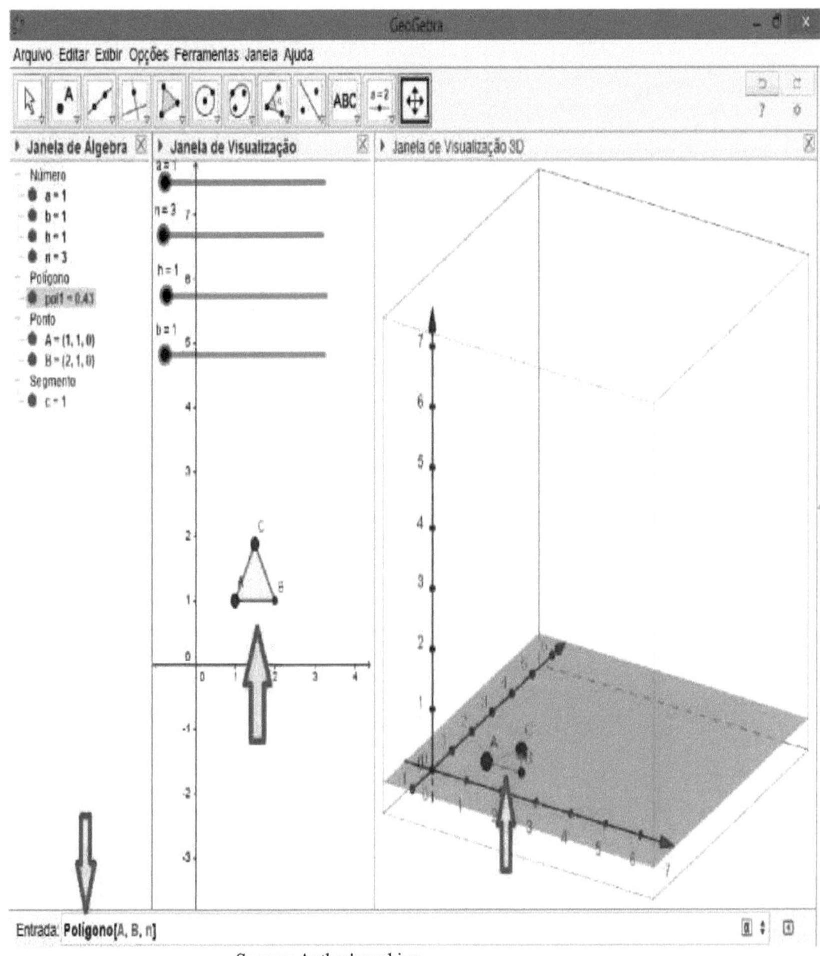

Source: Author's archive

Type the point "M=(b,l,h)", which will be the reference for the height of the prism, in "Input", and press *"Enter",* a point will be located in the vertical alignment of point "A" with distance "h", in the *3D* visualization window.

Figure 3.10: Locating point M

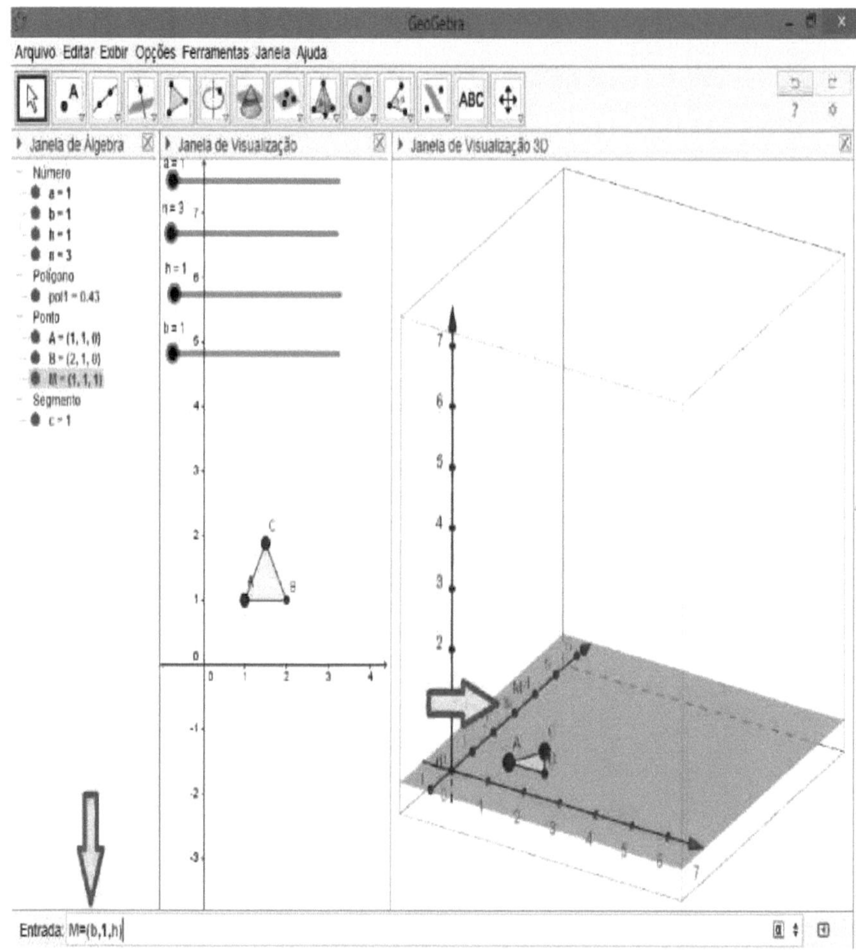

Source: Author's archive

Type the word "prism" in the "Input" bar and the following options will appear: "Prism[Polygon . Point]", select it and, under "polygon", type "poly" and, under "point", type "M": "Prism[Polygon . Point]", select it and, in "polygon", type "poly" and, in "point", type "M". When you press "*Enter"*, a regular triangular prism with a base side and height equal to 1 will appear.

Figure 3.11: Prism construction

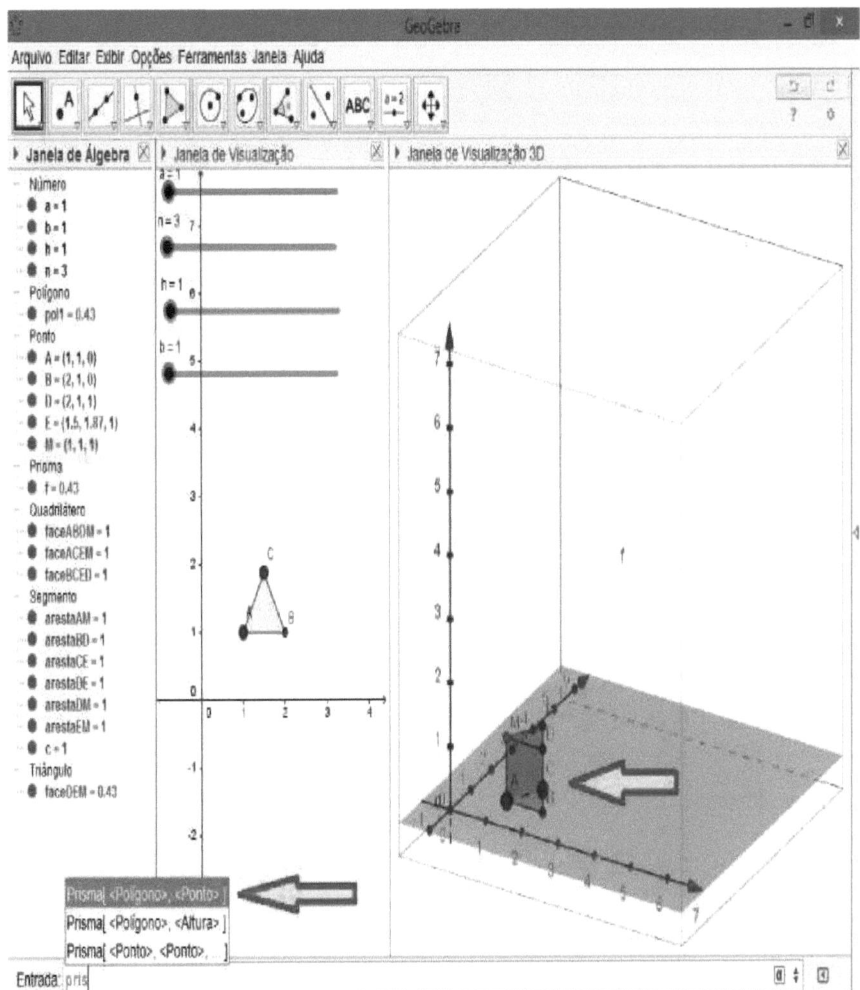

Source. Author's archivo

With the "3D visualization window" selected (1), the quick commands will show the drawing of an angle (2), click on the arrow in the bottom corner and on the right-hand side a number of options will appear, including the area calculation and the volumetric calculation. Select "volumc" (3) and click on the prism (4) and the volumc value of this solid will appear.

Figure 3.12: Calculating the volume of the prism

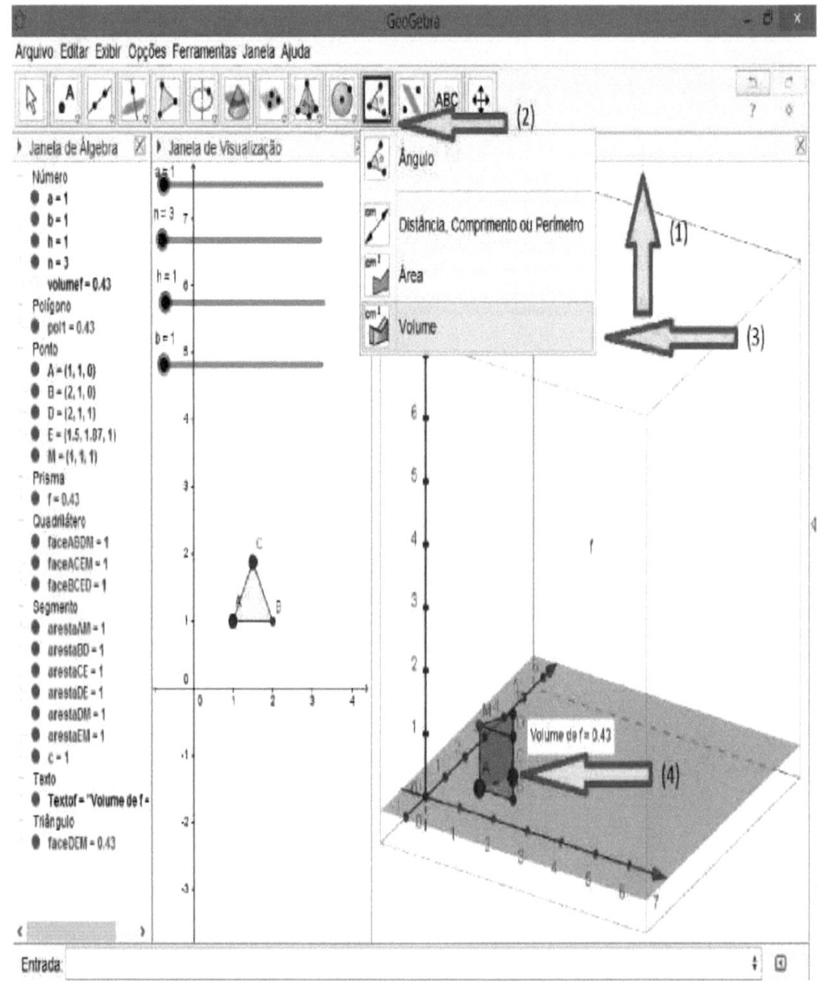

Source: Author's archive

Return to the same window (2) and select "area" (3) and, in the visualization window (1), click on the base polygon (4), and the area of the base will be shown.

Figure 3.13: Calculating the area of the base of the prism

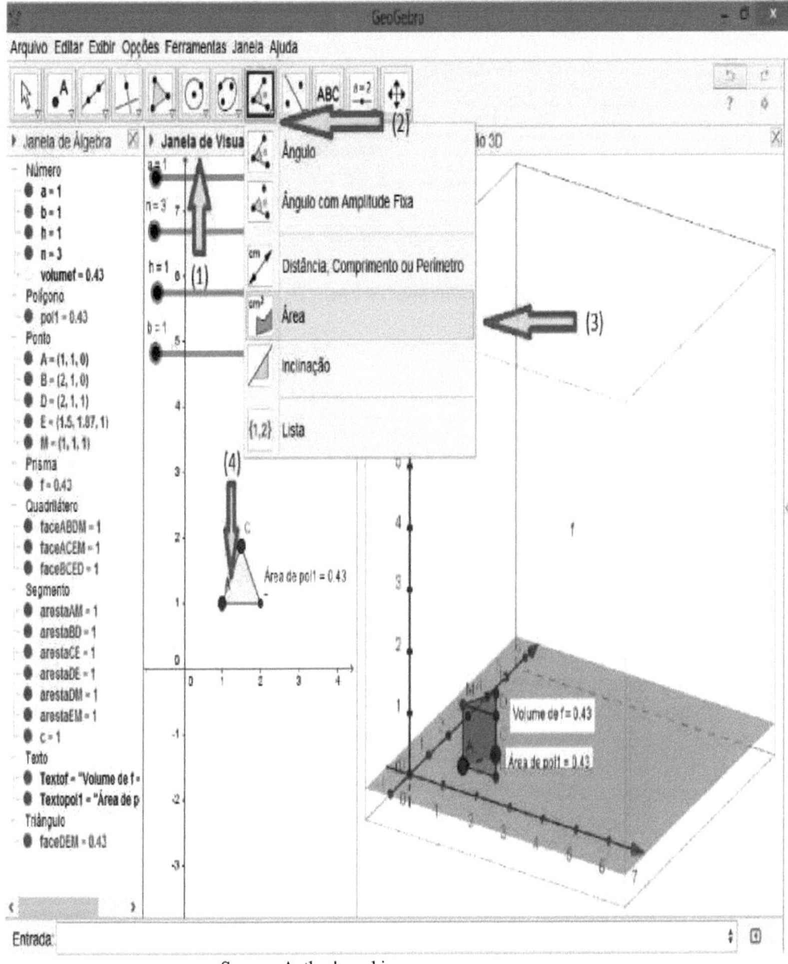

Source: Author's archive

With the *3D* visualization window selected (1), there will be a pyramid drawing, click on the arrow in the bottom right-hand corner (2), select "Planifying" (3) and click on the prism (4), the planification of the prism will be shown.

Figure 3.14: Planning the prism

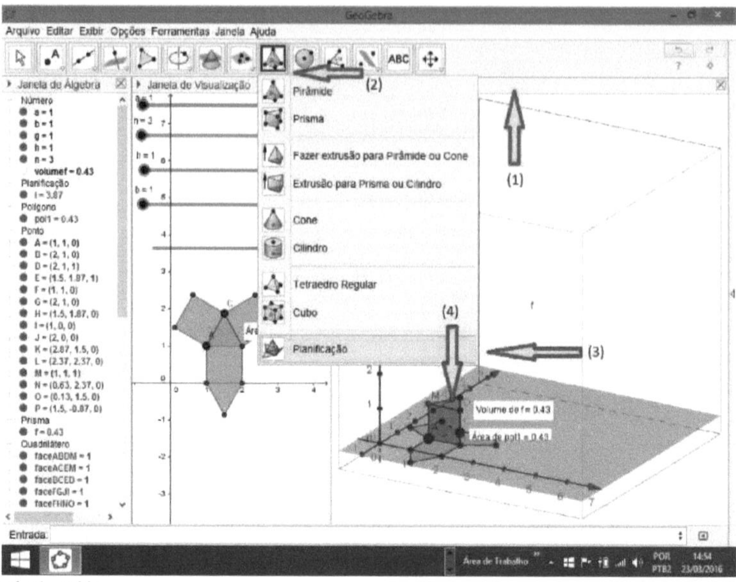

Source: Author's archive

The sliders are for modifying the prism, "a" changes the size of the side of the base, "n" defines the number of sides of the base, "h" 6 the height of the prism and "b" will show the prism straight, when "b=1", or oblique for other values.

When the value of "a" or "n" is changed, it can be shown that the volume and area of the base are the same when "h" is one.

Figure 3.15 Equality of the volume and area of the base of the prism when the height is one

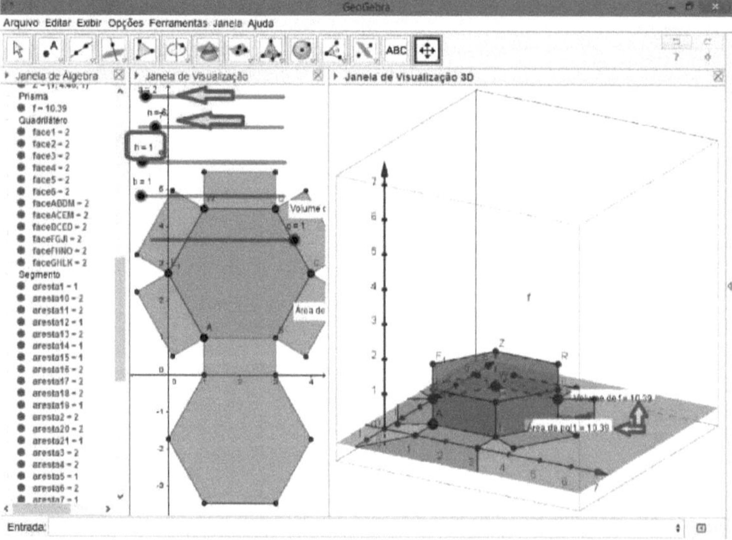

Source: Author's archive

If you change the value of "h", you'll notice that the volume is equal to h times the area of the base.

Figure 3.16 Comparison of volume and area of prism base as a function of height

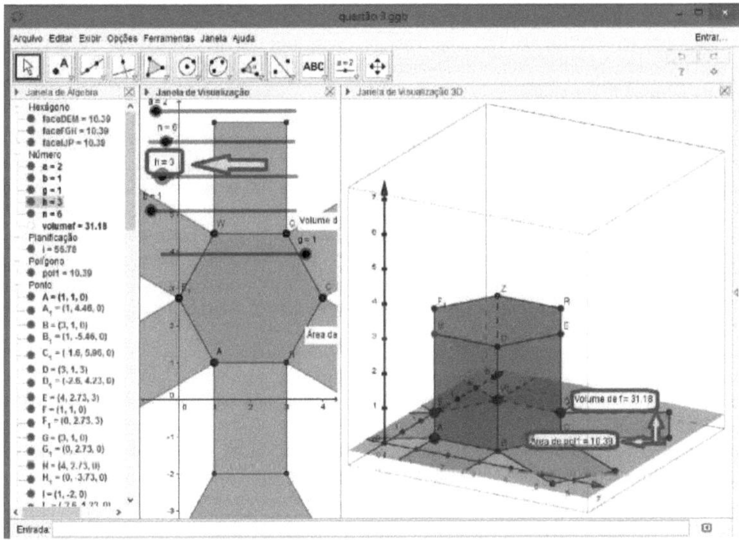

Source: Author's archive

By varying the value of "b", you can see that straight or oblique prisms with the same base and height have the same volume. In all these modifications, the heights can be shown on the plan of the prism.

Figure 3.17: Equal volume for straight and oblique prisms

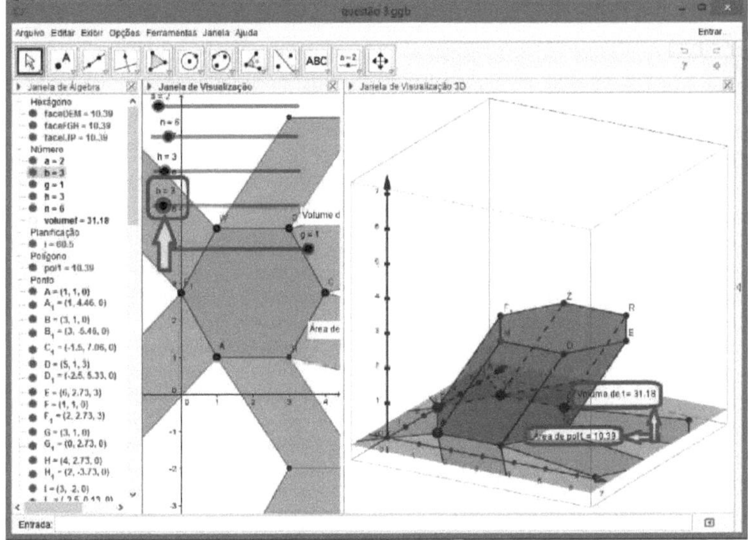

Source: Author's archive

For a perfect visualization of the planform, right-click on the "g slider" and enable "Animate". The *3D* window will show the shape of the solid and then its planform.

With this activity and their interest in computers, the students, encouraged by the teacher to build other solids, such as pyramids, cylinders and cones, which are made in the same way, but only the first one has a plan and side area determined by GeoGebra, will be able to discover an interest in learning Spatial Metric Geometry that will help them learn these concepts.

3.2.4 Fourth activity: main problem

The main objective of the activity is to work on calculating the area and volume of geometric solids such as prisms, pyramids, cylinders, cones and spheres. We will use concepts from Plane Geometry and animal diets, which are familiar to the students.

(1) Objective: to learn about geometric solids, their elements, areas and volumes;

(2) Reason: to understand geometric solids by determining how much material is needed to build them and what their capacity is;

(3) Description of the problem: A cattle farmer wants to feed 400 male Nelore cattle for 100 days. It is known that 7 kg of corn grain per animal per day will be used in the diet and that the density of corn grain is 1.244 ton/m^3 . You have been hired to indicate the shape and dimensions of the silo that will store this corn. You know that silos can be built in cylindrical, conical, spherical, pyramidal and prism shapes, at the same cost per square meter of building surface. Adopt a height equal to the diameter of the base in the case of cylinders and cones, and a height equal to the side of the base (square) in the case of pyramids and prisms.

   a) calculate the volume of corn to be ensiled;

   b) determine the dimensions of each solid;

   c) construct the plannings of these five solids, using a scale of 1:100, using cardboard for these plannings;

   d) with these templates, cut out the plannings in EVA and build a model of these solids;

   e) determine how much material was used to build each solid in the real situation;

   f) gives the ideal recommendation to the farmer as to the type of silo that should be built;

   g) comment on the feasibility of your appointment.

(4) Skills: Identify the elements of a prism, pyramid, cylinder, cone and sphere. Recognize the planning of common three-dimensional figures: cube, rectangular parallelepiped, straight prisms, pyramid, cylinder and cone. Solve problems involving the calculation of the lateral or total area of three-dimensional figures. Solve problems that involve calculating the volume of solids (SEEMG, p. 5);

(5) Technology: notebook, pencil, eraser, book, computer (GeoGebra), simple research using a cell phone or computer;

(6) Script: will be presented during the activity;

(7) Evaluation: check, through other questions, whether the student knows how to identify the solids, calculate the surface areas and their volumes, as well as the students' interest in developing the activity;

(8) Possible solution: follow the script together with the activity.

Nunes (2010, p. 94) cites that:

> In 1998, Onuchic elaborated some questions that can help teachers to reflect on them and to choose the problems they want to work on:
> (1) Is that a problem? Why?
> (2) What math topics can be started with this problem?
> (3) Is there a need to consider minor (secondary) problems associated with it?
> (4) What series do you think this problem is suitable for?
> (5) What paths could be taken to reach a solution?
> (6) How do you check the reasonableness of the answers you get?
> (7) As a teacher, would you find it difficult to work on this problem?
> (8) What degree of difficulty do you think your student might have with this problem?
> (9) How can the given problem be related to social and cultural aspects?
> (NUNES, 2010, p. 94)

For the proposed problem, we will answer the nine questions in order.

(1) Yes, according to Allevato (2005) and Onuchic (2004), a question is a problem if the students don't yet know the steps to solve it, but are interested in solving it. As this problem deals with concepts that may be required in their professions, they may be interested in solving it.

(2) This problem deals with the area and volume of geometric solids.

(3) Yes, we pointed out two secondary problems.

(4) The problem is suitable for the second year of high school.

(5) Determine the volume of the corn; calculate the dimensions of the solids; determine the area for building the solids; check the one with the lowest consumption.

(6) Observe whether the concepts of Spatial Metric Geometry have been applied consistently.

(7) No, because it's a sequential problem and well-defined in its objectives.

(8) This problem is of medium difficulty, as it deals with Plane Geometry concepts and leads to a search for formulas to apply.

(9) Use the solids determined to explain the feasibility of their construction and make the students aware of the types of silos that exist.

This fourth activity should be developed in four meetings, each lasting approximately 90 minutes, in which the three moments mentioned by Nunes (2010) should be observed: *before, during* and *after*.

First encounter: solving items (a) and (b).

In this first meeting, we will apply "the *before"*: by proposing that the students be divided into groups, preferably with three members in each; instructing the students on the procedure, the activity should be carried out by them independently, they can consult printed or virtual bibliographies and the teacher will observe their involvement, without assessing them for correctness, in carrying out the task; distributing the problem to the groups, handing out a copy of the problem to each group. From this moment on, we move on to the *during:* observing the students' posture; interfering only to encourage the appropriate conduct of research in the textbook adopted by the school or on search *sites*; pointing out the solutions, observing the paths taken to solve the problem, these spaces covered will be extremely important in the educator's final approach. For this meeting, the activities in letters (a) and (b) should be carried out.

During this meeting, the students will determine, in item (a), the weight at 280,000 Kg, which is equivalent to 225.08m³ of corn, but following a concept of technical reserve[7] for agricultural activity, the students will feel the need to add 10 percent to this volume, to meet possible unforeseen events, such as: not selling at 100 days or greater consumption by the animals, remembering that they cannot be left without food at any stage of the confinement. Therefore, this volume will rise to 247.59m³ ;

$$400 Kg * 100 * 7 = 280.000 Kg = 280 ton$$

$$280/1,244 = 225,08 m^3$$

$$225,08 * 1,10 = 247,59 m^3$$

In item (b), the dimensions determined, already with this addition, will be for the cylinder, the radius of the base of 3.40m, the height of 6.80m; for the cone, the radius of the base of 4.91m, the height of 9.82m; for the prism, the side of the base and the height of 6.28m; for the pyramid, the side of the base and the height of 9.06m; for the sphere, the radius of 3.90m.

$$\pi * R^2 * h \quad 247,59$$

$$(h \quad 2 * R)$$

$$\pi * R^2 * 2 * R = 247,59$$

Cylinder:
$$2 * \pi * R^3 \quad 247,59$$

$$R^3 = 39,41$$

$$R = 3,40 m$$

$$h = 2 * 3,40 = 6,80 m$$

Cone:

---

[7] It's an increase in the amount of food to cover some incident

$$\frac{1}{3}\pi * R^2 * h = 247,59$$

$$(h = 2 * R)$$

$$\frac{1}{3}\pi * R^2 * 2 * R = 247,59$$

$$\frac{2}{3} * \pi * R^3 = 247,59$$

$$R^3 = 118,22$$

$$R = 4,91m$$

$$h = 2 * 4,91 = 9,82m$$

Prisma:

$$l^2 * h = 247,59$$

$$(h = l)$$

$$l^2 * l = 247,59$$

$$l^3 = 247,59$$

$$l = 6,28m$$

$$h = 6,28m$$

Pyramid:

$$\frac{1}{3} * l^2 * h = 247,59$$

$$(h = l)$$

$$\frac{1}{3}l^2 * l = 247,59$$

$$\frac{1}{3} * l^3 = 247,59$$

$$l^3 = 742,77$$

$$l = 9,06m$$

$$h = 9,06m$$

Sphere:

$$\frac{4}{3}\pi * R^3 = 247,59$$

$$R^3 = 59,11$$

$$R = 3,90m$$

These values would be the correct ones for the problem, but remember that Problem Solving doesn't just value the correct answer but the whole way of learning, so the teacher can't influence everyone to get these values.

As an extra activity, encourage them to build these solids in GeoGebra and determine their volumes. Instruct them that the *software* doesn't calculate the areas of round surfaces, so that they don't lose track of this.

Second meeting: construction of geometric solids.

With the research materials they used in the first meeting, the students will see what the plannings of the solids look like, which will determine the dimensions, shapes and elements that are missing from their constructions, such as the generatrices of the cone and pyramid, as well as the central angle of the circular sector generated by the plannings of the cone. These elements must be discovered by the students according to the need they feel for this construction. These searches will lead to the conclusion that planifications are possible for the cylinder, the cone, the pyramid and the prism, but not for the sphere, as it does not have plannable surfaces.

To build the solids, the students will need cardboard, EVA, scissors, EVA glue or hot glue, a square, a protractor and a ruler.

Do the activity in class, first using the paper or cardboard and use these templates to cut out the EVA to build these four solids.

Third meeting: answering letters (e), (f) and (g)

The students will calculate the surface areas of the solids to check that the silo is viable. The calculations will show that the surfaces will have areas (St) of 217.90m^2 for the cylinder, 245.11m^2 for the cone, 236.63m^2 for the prism, 265.64m^2 for the pyramid and 191.13//^{2} for the sphere.

Cylinder:

$$St = 2 * Sb + Sl$$

$$St = 2 * \pi * R^2 + 2 * \pi * R * h$$

$$St = 2 * \pi * 3,40^2 + 2 * \pi * 3,40 * 6,80$$

$$St = 217,90m^2$$

Cone:

$$St \quad Sb + Sl$$

$$St = \pi * R^2 \mid \pi * R * g$$

$$(g^2 = R^2 \mid h^2) \Rightarrow (g^2 = 4,91^2 \mid 9.82^2) \Rightarrow (g = 10.98m)$$

$$St \quad \pi * 4,91^2 + \pi * 4,91 * 10,98$$

$$St - 245,11m^2$$

Prisma:

$$St = 2 * Sb + Sl$$

$$St = 2 * l^2 + 4 * l * h$$

$$St = 2 * 6,28^2 + 4 * 6,28 * 6,28$$

$$St = 236,63m^2$$

Pirámide-

$$St = Sb + Sl$$

$$St = l^2 + 4 * \frac{l * g}{2}$$

$$(g^2 = (\frac{l}{2})^2 + h^2) \Rightarrow (g^2 = (\frac{9,06}{2})^2 + 9,06^2) \Rightarrow (g = 10.13m)$$

$$St = 9,06^2 + 2 * 9,06 * 10,13$$

$$St = 265,64m^2$$

Sphere:

$$St = 4 * \pi * R^2$$

$$St = 4 * \pi * 3.90^2$$

$$St = 191.13m^2$$

Figure 3.18: Illustration of the answer to the fourth activity

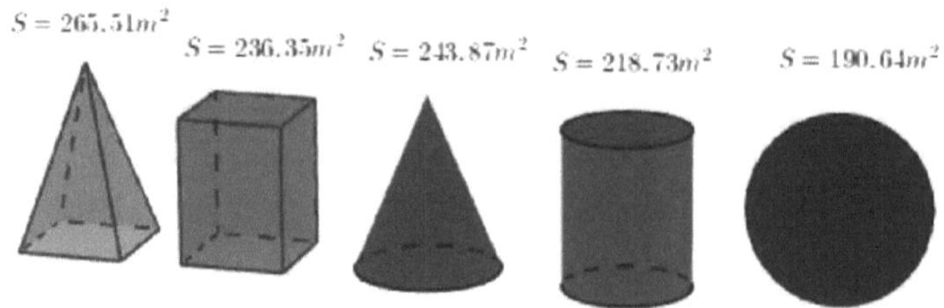

Source: Author's archive

Fourth meeting: the moment *after*.

Finally, when socializing the results, students should be encouraged to comment on what they have determined, the greatest difficulties, the best paths, the different shapes they have used, remembering that successes and errors are equally important, as the activity has built knowledge with successes but also with errors. Formalize the calculations of areas, planifications and volumes to conclude this activity and encourage them to do more research on the subject. Propose determining the area and volume ratios between the constructed solids (models) and the real ones. This proposal will work on the proportionality of similar solids, because when the solids are similar, the ratio between the areas is the square of the similarity ratio and the ratio between the volumes is the cube of the similarity ratio. With the above proposal, worked on by the teacher, the student will feel the need for further research. Nunes (2010) comments on the importance that *the after* activity plays in the teaching-learning process *through* Problem Solving, stating that learning does not end with solving the problem and that students should be encouraged to look for other ways of solving it.

The calculations in this fourth activity were carried out with the help of computerized resources, such as a calculator and GeoGebra *software,* to calculate the base and side areas of the prism and pyramid.

With this activity, we can work on the areas, volumes and plannings of the geometric solids, cylinder, cone, prism, pyramid and sphere, *through* Problem Solving, with a computational approach, using the vocational concepts of the students of the technical course in Agriculture and, with some adjustments, various vocational courses, as well as high school students in general. When an activity is planned *using* Problem Solving, it can arouse students' interest in learning mathematics in a broad and unrestricted way, so that students can see applications for the concepts involved in the problems. Brasil *apud* Allevato (2005, p. 22) states that "(...) for the history of science, we note that the problem invariably precedes the discoveries, is the provocateur of studies and the guide of theoretical constructions."

Final thoughts

Mathematics, because it is broad and has many regions, has a high degree of learning difficulty and is responsible for a large part of school retention in Brazil and around the world, at all levels of education. In order to change this reality in Brazil, the PCNs for Mathematics point to the use of Problem Solving as a starting point and highlight the importance of the History of Mathematics and Technologies as ways of learning Mathematics (BRASIL, 1998).

The teaching and learning of mathematics must be varied, and we have mentioned various trends in mathematics education. Using them in combination with pure mathematics, which in some cases will be indispensable, will make the teacher's task less arduous and more enjoyable. Creative lessons, with well-defined objectives, diversified ways of teaching, can increase student learning and thus generate respect for the teacher, rather than fear of the subjects, especially mathematics.

Problem solving can help students to become more interested in mathematics, as they will learn new concepts while already knowing at least one that is used in their daily lives. But the trends in teaching mathematics can't make all students learn - we know that no form of teaching can guarantee that everyone will learn.

This work proposes an activity that combines two teaching-learning trends, so we believe that if it is well explored by secondary school teachers, it will facilitate their work, since the technologies that students use in their daily lives, combined with problem-solving, can awaken their desire to learn. The teacher's job is an antithesis, mixing the ardor of teaching with the pleasure of learning.

The trends used in this dissertation, when worked on together, intertwine in such a way that it is difficult to ascertain which one is being used at any given time. Problem-solving builds on pre-acquired concepts to teach new ones, and technologies help in this construction at all times. Combining them creates a different way of working with mathematics, transforming them into a unique and effective teaching methodology.

The teacher's creativity in conducting his or her lessons, with the use of different teaching-learning methodologies, can result in more engaging lessons, with students who are more reflective and critical in their construction of knowledge.

REFERENCES

ALLEVATO, N. S. G. *Associando o computador á resolugáo de problemas fechados:* análise de urna experiencia, institute of geosciences and exact sciences, Rio Claro, 2005.

ARAÚJO, .1. 1.: PINTO, M. M. F.; LUZ, C. R. da; RIBEIRO, A. R. *Ephemeralities of scenarios for investigation in a mathematics classroom episode with technologies.* Zetetiké, Campiñas, v. 16 n. 29, jan/jun. 2008, p. 7 - 40.

ARAÚJO, P. B. *Situagoes de aprendizagem: a circunferencia, a mediatriz e urna abordagem am 2 GeoCebra.* Sao Paulo, 2010

BASSANEZI, R. **C. Mathematical modeling - an emerging discipline in teacher training programs.** In: XXII National Congress of Applied and Computational Mathematics, 1999, Santos. Biomathematics IX. Campiñas: IMECC, 1999. v. 9. p. 9-22.

BRAZIL. Ministry of Education. Secretariat for Primary Education. *Parámetros curriculares nacionais* - Matemática: terceiro e quarto ciclos. Brasilia: MEC/SEF, 1998, 148p.

BORBA, M. C. Will information technology bring changes to Brazilian education? Zetetiké, Campiñas, v. 4, n. 6, jul/dez. 1996, p. 123 - 134.

COSTA, F. J. M. da. *Ethnomathematics:* methodology, tool or simply ethnorevolution? Zetetiké, Campiñas, v. 22, n. 42, jul/dez. 2014, p. 181 - 196.

DAMACENO, D. S. *A Resolugáo de Problemas e os aspectos significativos da* sua praíwa *n0s* "idos de *Matemática,* VI EPCT, Campo Mouráo, 2011.

DAVID, M. M. M. S. *As possibilidades de inovagáo no ensino-aprendizagem* da *Matemática dementar.* Presenta Pedagógica, Belo Horizonte, v. 1, n. 1, jan./feb. 1995, p.56 - 66.

FIORENTINI, D. e MIORIM, M. A. *Urna reflexáo sobre o uso de materiais concretos e jogos no Ensino da Matemática.* Published in SBEM-SP Bulletin, Year 4, n. 7, 1996, p. 1 - 4.

FLEMMING, D. M.; LUZ, E. F.: MELLO, A. C. C. de. *Tendencia em Educagáo Matemática:* Disciplina na modalidade á distancia. 2ª ediQáo - PalhoQa: UnisulVirtual, 2005, 87 p.

FREDERICO, F. T.; GIANOTO, D. E. P. *Teaching science and mathematics: the* use of information technology and teacher training. Zetetiké, Campiñas, v. 22, n. 42, jul/dez. 2014, p. 63 - 88.

GOMES, T. de A.; RODRIGUES, C. K. *The Evolution of Trends in Mathematics Education and the Approach of the History of Mathematics in Teaching.* Revista de Educagáo, Ciencias e Matemática v.4 n.3, Sep/Dec 2014, p. 57-67.

MENDES, I. A. *Matemática em sala de aula: tecendo* redes cognitivas na aprendizagem. Natal: Flecha do tempo, 2006, 120 p.

MIGUEL, A. The pedagogical potential of the History of Mathematics in question: reforming and questioning arguments. **Zetetiké,** Campiñas, v. 5, n. 8, p. 73-105, Jul./Dec. 1997

MISKULIN, R. G. S. *Reflexdes sobre as tendencias actuais da Educagáo Matemática e da informática,* cempem. unicamp, tese de doutororado. 1999, 32 p.

MISKULIN, R. G. S. *Identification and analysis of the dimensions that permeate the use of information and communication technologies in mathematics classes in the context of teacher training.* Bolema, Rio Claro, v. 19, n. 26, 2006, p. 103 - 123.

MURARI, C. *Experiencing manipulative materials for teaching* and *learning mathematics.* Bolema, Rio Claro, v. 25, n. 41, dec. 2011, p. 187 - 211.

NUNES, C.B. *The Teaching-Learning-Evaluation Process of Geometry through Problem Solving:* didactic-mathematical perspectives in the initial training of mathematics teachers. Rio Claro, 2010. 337 p.

ONUCHIC, L. de la R.; ALLEVATO, N. S. G. *New reflections on teaching and learning mathematics through problem solving.* In: BICUDO, M. A. V.; BORBA, M. de C. (Org.). *Educagáo Matemática:* pesquisa em movimento. Sao Paulo: Cortez, 2004, p. 213 - 231.

ONUCHIC, L. de la R. *Problem solving in mathematics education:* where are we? And where are we going? Espado pedagógico, Passo Fundo, v. 20, n. 1, Jan/jun 2013, p.88 - 104.

PENTEADO, M. G. *Redes de Trabalhos:* Expansão das Possibilidades da Informática na Educagáo Matemática da Escola Básica. In: BICUDO, M. A. V.; BORBA, M. de C. (Org.). Educagáo Matemática: pesquisa em movimento. São Paulo: Cortez, 2004, p. 283 - 295.

PEREIRA, G. H. A. *Ethnomathematics as the valorization of disc8nt knowledge:* context and meaning in Mathematics Education. 2008. 60 f.

Monograph (Degree in Mathematics) - Faculty of Philosophy, Sciences and Letters of Alto São Francisco, Luz, 2008.

PINTO, N. B. *Historical marks of modern mathematics in Brazil.* Revista Diálogo Educacional, Curitiba, v.5, n.16, 2005.

POLYA, G. *The art of problem solving: a new aspect of the mathematical method.* Translated and adapted by Heitor Lisboa de Araujo. Rio de Janeiro: interciéncia, 1995, 196 p.

SEEMG, Minas Gerais State Department of Education ? Undersecretariat for the Development of Basic Education. *Common Basic Content (CBC)* for *Secondary School Mathematics 2 Supplementary* Meats 2013

SKOSVMOSE, Ole. *Critical Mathematics Education:* The Question of Democracy, Campinas, Papirus 2001, 160 p.

SILVA, E. L. da; MENEZES, E. M. *Metodología da pesquisa e elaboragáo de dissertáo.* 3. ed. rev. atual. Florianópolis: UFSC Distance Learning Laboratory, 2001, 121 p.

SIQUEIRA. R. A. N. de. *Tendencias da Educagáo Matemática* na *Formagáo* de *Pr0fi:ss0res.* Ponta Grossa, 2007, 50 p.

SKOVSMOSE, O. *Critical Mathematics Education:* The question of democracy. Campinas, SP. Papirus, 2001. 148 p.

SOARES. D. A. *Educagáo Matemática Crítica:* ContribuiQÓes para o Debate Teórico e sens Reflexos nos Trabalhos Acadêmicos. Sao Paulo: PUC 2008, 157 p.

VALENTE, J. A. *Different uses of computers in education.* Campinas, SP. Gráfica Central da UNICAMP, 1993 p. 1 - 23.

ZORZAN, A. S. L. *Teaching-learning:* some trends in Mathematics Education. R. Ciencias Humanas, Frederico Westphalen, v. 8, n. 10, Jun 2007, p. 77 - 93.

I want morebooks!

Buy your books fast and straightforward online - at one of world's fastest growing online book stores! Environmentally sound due to Print-on-Demand technologies.

Buy your books online at
www.morebooks.shop

Kaufen Sie Ihre Bücher schnell und unkompliziert online – auf einer der am schnellsten wachsenden Buchhandelsplattformen weltweit! Dank Print-On-Demand umwelt- und ressourcenschonend produziert.

Bücher schneller online kaufen
www.morebooks.shop

info@omniscriptum.com
www.omniscriptum.com

Printed by Books on Demand GmbH, Norderstedt / Germany